Immobilized Biocatalysts
Saccharomyces Yeasts
Wastewater Treatment

S. Fukui, A. Tanaka

Application of Biocatalysts Immobilized by Prepolymer Methods

T. Kamihara, I. Nakamura

Regulation of Respiration and Its Related Metabolism by Vitamin B_1 and Vitamin B_6 in *Saccharomyces* Yeasts

H. Sahm

Anaerobic Wastewater Treatment

R. Sudo, S. Aiba

Role and Function of Protozoa in the Biological Treatment of Polluted Waters

THEORIA CUM PRAXI

AKADEMIE-VERLAG · BERLIN

Immobilized Biocatalysts
Saccharomyces Yeasts
Wastewater Treatment

Immobilized Biocatalysts *Saccharomyces* Yeasts Wastewater Treatment

Managing Editor: A. Fiechter

with 73 Figures and 45 Tables

Akademie-Verlag · Berlin 1984

Die Originalausgabe erscheint im Springer-Verlag
Berlin—Heidelberg—New York—Tokyo
als Volume 29 der Schriftenreihe Advances in
Biochemical Engineering/Biotechnology

Vertrieb ausschließlich für die DDR und die sozialistischen Länder

Erschienen im Akademie-Verlag Berlin, DDR-1086 Berlin, Leipziger Straße 3—4
Lizenznummer: 202 · 100/509/84
Printed in the German Democratic Republic
Gesamtherstellung: VEB Druckerei „Thomas Müntzer", 5820 Bad Langensalza
Umschlaggestaltung: Karl Salzbrunn
LSV 1315
Bestellnummer: 763 355 0 (6829)

07800

Table of Contents

Application of Biocatalysts Immobilized by Prepolymer Methods

Saburo Fukui and Atsuo Tanaka
Laboratory of Industrial Biochemistry, Department of Industrial Chemistry, Faculty of Engineering, Kyoto University, Yoshida, Sakyo-ku, Kyoto 606, Japan

Of the various immobilization methods available at present, entrapment of biocatalysts in natural or synthetic polymer gels is the most promising because of its applicability to the immobilization not only of single enzymes but also of plural enzymes, cellular organelles, microbial cells, plant cells and animal cells. This review summarizes novel and convenient methods of entrapping biocatalysts by using synthetic prepolymers of different chemical structures and physico-chemical properties. Emphasis is placed on the effects of the physico-chemical properties of gels prepared from various prepolymers which have different properties — net-work size, hydrophilicity-hydrophobicity balance and ionic nature — on the catalytic activities of the thus-entrapped biocatalysts. Bioconversions of hydrophobic or water-insoluble compounds by entrapped biocatalysts under hydrophobic conditions are also mentioned.

1 Introduction

Immobilization of biocatalysts — enzymes, cellular organelles, microbial cells, plant cells and animal cells — is attracting worldwide attention. Immobilized biocatalysts are, in general, stable and easy to handle compared with native counterparts. One of the most important features is that they can be used repeatedly in a long-term series of batch reactions or continuously in flow systems. At present, applications of immobilized biocatalysts include

(1) production of useful compounds by stereospecific and/or regiospecific bioconversion,
(2) production of energy and reducing equivalent by biological processes,
(3) selective treatment of specified pollutants to solve environmental problems,
(4) continuous analyses of various compounds with high sensitivity and high specificity, and
(5) utilization in medical fields, such as new types of drugs, artificial organs etc.

These processes require immobilization not only of single enzymes that catalyze simple reactions but also of plural numbers of enzymes or multi enzyme systems that mediate more complicated reactions.

Various immobilization techniques are available at present. The principle of these methods are:

(1) carrier binding via covalent linkage, electrostatic force or physical adsorption,
(2) crosslinking by using bi- or multi-functional reagents,
(3) entrapment in gel matrices, microcapsules, liposomes or hollow-fibers, and
(4) combination of the above-mentioned techniques [1,2].

Of these different methods developed hitherto, entrapment methods are the most promising because the technique is applicable to the immobilization of a variety of biocatalysts.

Different types of gel materials, such as polysaccharides, proteins and synthetic polymers, are now used to entrap biocatalysts. Polysaccharides and proteins have been proved to be useful. However, applications of immobilized biocatalysts are being aimed toward a wide variety of bioreactions, including synthesis, transformation, degradation or assay of various compounds having different chemical properties [3]. To satisfy different types of demands in such a variety of applications, it will be desirable to entrap biocatalysts in gels of adequate physico-chemical properties. It is not easy to select suitable gels for each purpose among those prepared from natural polymers. Furthermore, it will be rather difficult to modify polysaccharides and proteins for the preparation of proper gel materials.

Acrylamide and its analogues have been used widely as starting materials for gels. However, these monomers are sometimes liable to deactivate biocatalysts when the two are mixed together.

From these reasons continuous efforts have been devoted to developing new entrapment methods using synthetic prepolymers. This article describes mainly "prepolymer methods" which have been developed in the authors' laboratory and proved to be useful for various purposes.

2 Prepolymer Methods for Entrapment of Biocatalysts

Specific features and advantages of the prepolymer methods can be summarized as follows:

(1) Entrapment procedures are very simple under very mild conditions.
(2) Prepolymers do not contain monomers which may have bad effects on the biocatalysts to be entrapped.
(3) The net-work structure of gels can be controlled by using prepolymers of optional chain-length.
(4) The physico-chemical properties of gels, such as hydrophilicity-hydrophobicity balance and ionic nature, can be changed by selecting suitable prepolymers which were synthesized chemically in the absence of biocatalysts.

Several examples of the prepolymer methods will be described below.

2.1 Photo-crosslinkable Resin Prepolymer Method

Entrapment of biocatalysts should be carried out under mild conditions in which high temperatures, shifts of pH to extremely alkaline or acidic sides, etc. are avoided. In the case of photo-crosslinkable resin prepolymers, illumination with near-ultraviolet light initiates radical polymerization of the prepolymers and completes gel formation within only 3–5 minutes.

Various types of prepolymers having photosensitive functional groups have been developed by the present authors. The structures of typical photo-crosslinkable resin prepolymers are shown in Fig. 1, and the properties of several prepolymers are summarized in Table 1.

Poly(ethylene glycol) dimethacrylate (PEGM) was synthesized from poly(ethylene glycol) and methacrylate. ENT and ENTP were prepared from hydroxyethylacrylate,

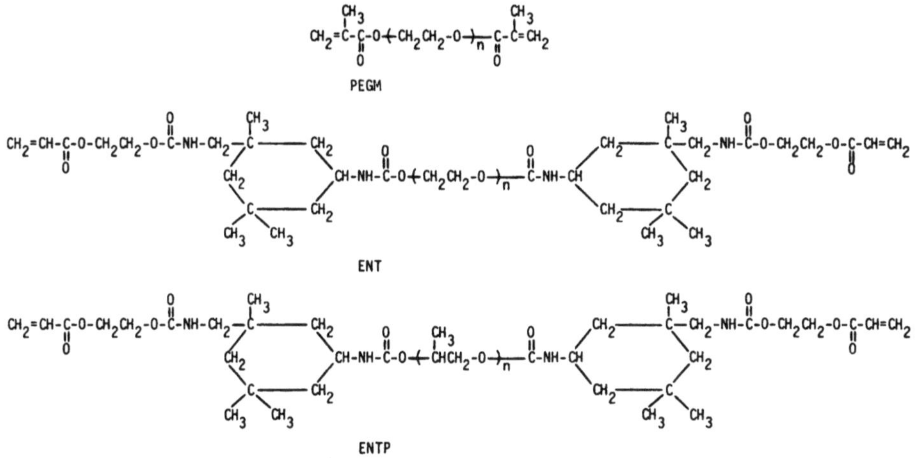

Fig. 1. Structures of typical photo-crosslinkable resin prepolymers. PEGM and ENT, water-soluble; ENTP, water-insoluble

Table 1. Properties of several photo-crosslinkable resin prepolymers

Prepolymer	Main chain	Mw of main chain	Property
PEGM-1000	Poly(ethylene glycol)	ca. 1,000	Hydrophilic
PEGM-2000		2,000	
PEGM-4000		4,000	
ENT-1000	Poly(ethylene glycol)	1,000	Hydrophilic
ENT-2000		2,000	
ENT-4000		4,000	
ENT-6000		6,000	
ENTP-1000	Poly(propylene glycol)	1,000	Hydrophobic
ENTP-2000		2,000	
ENTP-3000		3,000	
ENTP-4000		4,000	
ENTB-1000	Polybutadiene	1,000	Hydrophobic
PB-200k	Polybutadiene	200,000	Hydrophobic
PBM-2000	Maleic polybutadiene	9,000	Hydrophobic
ENTE-1	[Poly(vinyl alcohol)]	—	Emulsion-type
ENTA-1	Trimellitic poly(propylene glycol)	2,200	Anionic

isophorone diisocyanate and poly(ethylene glycol) or poly(propylene glycol), respectively. Each prepolymer has a linear skeleton of optional length, at the ends of which are attached the photosensitive functional groups, acryloyl, methacryloyl etc. PEGM [4-6] and ENT [7,8] prepolymers containing poly(ethylene glycol) as the main skeleton are hydrophilic and give hydrophilic gels, while ENTP [9] with poly(propylene glycol) as the main skeleton have a hydrophobic property and form hydrophobic gels. By using poly(ethylene glycol) or poly(propylene glycol) of different molecular weight, we can prepare prepolymers of different chain-length, that is, PEGM-1000 to PEGM-4000, ENT-1000 to ENT-6000, and ENTP-1000 to ENTP-4000. Chain-length of the prepolymers correlates to the size of net-work of the gels formed from these prepolymers. Anionic and cationic prepolymers can also be prepared by introducing anionic and cationic functional group(s) respectively to the main skeleton of prepolymers.

Typical immobilization procedures with photo-crosslinkable resin prepolymers are shown in Table 2. Upon illumination with near-ultraviolet light for several minutes,

Table 2. Typical examples of immobilization of bacterial cells with photo-crosslinkable resin prepolymers

Hydrophilic prepolymer		Hydrophobic prepolymer	
ENT-4000	1.0 g	ENTP-2000	1.0 g
Benzoin ethyl ether	10 mg	Benzoin ethyl ether	10 mg
2 mM Potassium phosphate buffer, pH 7.0	3 ml	Benzene-*n*-Heptane (1:1 by volume)	3 ml
Wet cells	1.0 g	Wet cells	1.0 g

Illumination of near-ultraviolet light (wavelength range, 300–400 nm; maximum intensity at 360 nm) for 3 min the photosensitive groups of the prepolymers become free radicals which quickly crosslink with each other. Gels entrapping biocatalysts can be obtained by illumination of a mixture consisting of a prepolymer, a photo-sensitizer such as benzoin ethyl ether or benzoin isobutyl ether, and a solution or suspension of biocatalysts. A suitable buffer is used for the hydrophilic prepolymer and an adequate organic solvent for the hydrophobic prepolymer [10]. Suitable mixtures of these two types of prepolymers can also be utilized [9]. In some cases, a detergent is employed to mix a hydrophobic prepolymer with an aqueous solution or suspension of biocatalysts [11]. These prepolymers have been applied for entrapment not only of enzymes but also of microbial cells and cellular organelles [12, 13].

An emulsion-type photo-crosslinkable resin prepolymer is useful when swelling of the gels is to be avoided [14]. ENTE-1 was prepared as follows: Hydrophobic poly(vinyl acetate) was coated with hydrophilic poly(vinyl alcohol) to be readily dispersed in an aqueous solution. Photosensitivity was introduced to the particles by reacting with *N*-hydroxymethylacrylamide in the presence of an acid catalyst. After neutralization, the resulting prepolymer was used as an emulsion-type photo-crosslinkable resin prepolymer.

Immobilized enzyme tubes or microbial cell tubes can be easily prepared by using ENTE-1. The inner surface of small glass tubes is silanized with vinyltriethoxysilane. A mixture of ENTE-1, photosensitizer and a biocatalyst is injected at the top of each silanized glass tube, forming a thin layer of the biocatalyst-containing prepolymer on the inner surface of the tube. Each tube is illuminated with near-ultraviolet light for several minutes under an atmosphere of nitrogen to complete the photo-crosslinking of the prepolymer. ENTE-1 has a double-stratified structure (core, hydrophobic poly(vinyl acetate); outer layer, hydrophilic poly(vinyl alcohol)). The structure of the prepolymer prevents the resulting gel from swelling in an aqueous solution and enables biocatalyst-entrappling films to be attached tightly to the glass surface compared with other photo-crosslinked resins. The resulting biocatalyst tubes are useful for the continuous assay of various compounds.

As described above, different types of photo-crosslinkable resin prepolymers can be used according to the purposes.

Poly(vinyl alcohol) bearing aromatic azido groups was also applied for immobilization of an enzyme. In this case, longer illumination of near-ultraviolet light was necessary to complete the polymerization [15].

2.2 Urethane Prepolymer Method

Entrapment of biocatalysts with water-miscible urethane prepolymers (Fig. 2) seems to be most simple and convenient, because isocyanate functional groups at both terminals of the molecule react with each other only in the presence of water, forming a urea linkage and liberating carbon dioxide [16]. That is, when the prepolymers are mixed with an aqueous solution or an aqueous suspension of a biocatalyst, gels are easily formed within a few minutes and gelation is complete in 30 to 60 min. Prepolymers with different hydrophilic or hydrophobic character can be obtained by changing the ratio of the poly(ethylene glycol) part and poly(propylene glycol) part in the polyether diol moiety of the prepolymers (Table 3). For example, PU-3 with a high content of poly(propylene glycol) gives hydrophobic gels, while PU-6 and PU-9

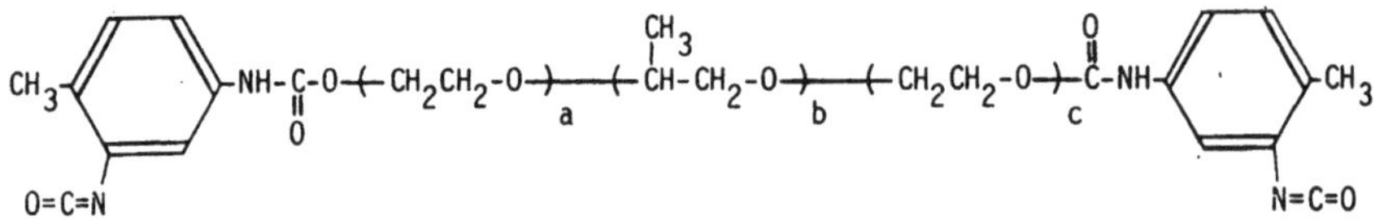

Fig. 2. General formula of water-miscible urethane prepolymer (PU)

with a high content of poly(ethylene glycol) give hydrophilic gels, although all the prepolymers are water-miscible [13,17]. The chain-length and the content of isocyanate group can also be changed. These urethane prepolymers can also be used for the entrapment of various biocatalysts, such as enzymes, microbial cells and cellular organelles [13,18].

Table 3. Properties of urethane prepolymers [17]

Prepolymer	Mw of polyether diol	NCO content in prepolymer (%)	Poly(ethylene glycol) content in polyether diol (%)
PU-1	1,486	4.0	100
PU-2	2,529	3.1	57
PU-3	2,529	4.2	57
PU-4	2,529	5.6	57
PU-5	2,627	2.7	91
PU-6	2,627	4.0	91
PU-7	2,627	5.6	91
PU-8	2,616	2.7	100
PU-9	2,616	4.0	100
PU-10	2,616	5.6	100
PU-11	4,285	4.0	73

Similar type of prepolymers available commercially have also been used for entrapment of microbial cells [19,20].

2.3 Radiation Method

Various types of functional monomers have been employed for the entrapment of biocatalysts by γ-irradiation. Several prepolymers, such as poly(vinyl alcohol) [21], poly(ethylene glycol) diacrylate [22] and poly(ethylene glycol) dimethacrylate [22,23], are also useful gel materials for entrapment by irradiation.

When 10% of poly(vinyl alcohol) was used, 5–7 Mrad of irradiation was sufficient to obtain rigid gels [21]. Kaetsu and his coworkers developed a technique for immobilizing biocatalysts by means of radiation polymerization of glass-forming prepolymers, such as poly(ethylene glycol) diacrylate and poly(ethylene glycol) dimethacrylate, at low temperature. Entrapment was carried out at −24 °C with 10% of a prepolymer and an irradiation of 1 Mrad [23]. A mixture of poly(ethylene glycol) dimethacrylate and poly(vinyl alcohol) was also applicable [22].

The radiation method has several advantages such as possible application of a wide variety of monomers and prepolymers as gel materials and immobilization at low temperature. However, utilization of the method is limited because radiation apparatus is required.

2.4 Miscellaneous

Owing to the advantages of the prepolymer methods mentioned above, various types of synthetic resin prepolymers have been proposed.

Freeman and Aharonowitz [24,25] synthesized water-soluble linear polyacrylamide (Mw, $15\text{–}17 \times 10^4$) partially substituted with the acylhydrazide group. This prepolymer can be gelled in the presence of a crosslinking agent such as glyoxal, glutaraldehyde or periodate-oxidized poly(vinyl alcohol).

Poly(acrylamide-co-*N*-acryloxysuccinimide) (PAN) was synthesized from acrylamide, *N*-acryloxysuccinimide and azobis-isobutyronitrile. Gel formation and enzyme immobilization were accomplished simultaneously by the reaction of PAN with triethylenetetramine (crosslinking agent) and an enzyme. In this case, the enzyme protein is bound covalently to the polymer through an amide-forming reaction [26].

Poly(ethylene glycol) dimethacrylate was also used as a crosslinking agent for the entrapment of biocatalysts by radical polymerization of acrylic acid and *N,N*-dimethylaminoethyl methacrylate [27].

Application of epoxy resins [28] and methacrylic ester copolymers [29] were also reported. Prepolymers having different functional groups are interesting and important for immobilizing biocatalysts by different types of reactions.

3 Biocatalysts Entrapped by Prepolymer Methods

As described above, various kinds of prepolymers having different physico-chemical properties have been developed for the immobilization not only of enzymes but also of microbial cells (dead or alive) and very unstable organelles.

3.1 Enzymes

Enzymes immobilized with prepolymers are listed in Table 4. Several enzymes, such as invertase [4,5,11,16,21,26], catalase [8,11], glucoamylase [21] and β-glucosidase [15], were used as test enzymes to develop new immobilization techniques. Only a few reports are available concerning the applications of enzymes immobilized by prepolymer methods.

Baughn et al. [30] examined the large-scale production of ATP from adenosine and acetyl phosphate by a combination of adenosine kinase, adenylate kinase and acetate kinase immobilized with poly(acrylamide-co-*N*-acryloxysuccinimide) (PAN). From 150 mmol of adenosine, 125 mmol of ATP, 20 mmol of ADP and 3 mmol of AMP were synthesized. The yield of phosphorylated adenosine derivatives was 38 %, based on the acetyl phosphate added. In this case, the mixed enzyme system was preferable for the sequential reaction system. Furthermore, adenosine kinase and acetate kinase should be co-immobilized in order to catalyze the phosphorylation of adenosine with acetyl phosphate effectively.

Table 4. Enzymes immobilized by prepolymer methods

Enzyme	Immobilization method (prepolymer)	Application	Ref.
Invertase	Photo-crosslinking (PEGM *etc.*)	Hydrolysis of sucrose	4,5,11)
Invertase	Urea linkage (PU)	Hydrolysis of sucrose	16)
Catalase	Photo-crosslinking (ENT *etc.*)	Degradation of hydrogen peroxide	8,11)
Glucose isomerase	Photo-crosslinking (ENT)	Isomerization of glucose	13)
Amino acylase	Photo-crosslinking (ENT)	Resolution of DL-amino acids	13)
Trypsin, invertase *etc.* (20 enzymes)	Crosslinking with triethylenetetramine (PAN)	—	26)
Adenosine kinase + adenylate kinase + acetate kinase	Crosslinking with triethylenetetramine (PAN)	Production of ATP	30)
Creatine kinase + phosphofructokinase	Radical polymerization (PEGM + acrylic acid + *N*,*N*-dimethylaminoethyl methacrylate)	Production of ATP	27)
Lipase	Photo-crosslinking (ENTP)	Ester exchange of triglyceride	31,32)
Lipase	Photo-crosslinking (ENTP)	Hydrolysis of triglyceride	33)
Hydrogenase	Photo-crosslinking (ENT)	Reduction of NAD	34)

ENT, hydrophilic photo-crosslinkable resin prepolymer; ENTP, hydrophobic photo-crosslinkable resin prepolymer; PAN, poly(acrylamide-co-*N*-acryloxysuccinimide); PEGM, poly(ethylene glycol) dimethacrylate; PU, urethane prepolymer

Lipases entrapped with hydrophobic photo-crosslinkable resin prepolymers (ENTP-2000 and ENTP-4000) have been applied to the ester exchange reaction and hydrolysis of triacylglyceride. Production of cacao butter-like fat from olive oil and stearic acid or palmitic acid by enzymatic interesterification has been successfully achieved with gel-entrapped *Rhizopus delemar* lipase in an organic solvent system [31,32]. Hydrophobic prepolymers (ENTP) were found to be far superior to hydrophilic photo-crosslinkable resin prepolymers (ENT). Entrapment markedly enhanced the operational stability of the enzyme. In the case of the hydrolysis of olive oil to glycerol and fatty acids, *Candida cylindracea* lipase entrapped with hydrophobic prepolymers (ENTP) showed the highest activity among the immobilized preparations examined [33]. Entrapment also made it possible to use the enzyme repeatedly or continuously.

Enzymes entrapped with different photo-crosslinkable resin prepolymers were applied to the analyses of various compounds (Table 5). Enzyme tubes entrapping glutamate decarboxylase [14] and lysine decarboxylase [35] were prepared with the emulsion-type photo-crosslinkable resin prepolymer (ENTE-1). These enzyme tubes were used in continuous flow systems to measure the concentrations of L-glutamate and L-lysine, respectively, in fermentation broths. Thin enzyme films mounted on an oxygen electrode were also used as enzyme electrodes [13]. Several oxygen-consuming oxidases were employed for this purpose. These enzyme electrodes were found to be very stable.

Table 5. Several properties of enzyme electrodes and enzyme tubes prepared by photo-crosslinkable resin prepolymer methods [13]

Enzyme	Prepolymer(s) used	Substrate	Measurable range	Stability	Ref.
Choline oxidase	ENT-1000 + ENT-2000	Choline HCl	<5 mM	>300 assay (>86 days)	13)
	ENT-1000 + ENT-2000	Lecithin (+ phospholipase D)	~10 mM	>350 assay (>16 days)	
	ENT-1000 + ENT-2000 - ENTA	Choline HCl	<2 mM	>150 assay	
Glucose oxidase	ENT-4000	Glucose	~500 mg/dl	>100 assay	13)
Uricase	ENT-4000	Uric acid	5–50 mg/dl	>80 assay	13)
Lactate oxidase	ENT-2000	*DL*-Lactic acid	~50 mM	>35 assay	13)
Pyruvate oxidase	ENT-2000	Potassium pyruvate	~25 mM	>50 assay	13)
Glycerophosphate oxidase	ENT-2000	*DL*-α-glycerol-3-phosphate	~150 mM	>50 assay	13)
Glycerokinase + glycerophosphate oxidase	ENT-2000	Glycerol	~25 mM	>110 assay	13)
Glutamate decarboxylase	ENTE-1	*L*-Glutamic acid	<30 mM	>500 assay (>19 days)	14)
		Fermented broth	—	>70 assay	
Lysine decarboxylase	ENTE-1	*L*-Lysine	<2 mM	>500 assay	35)
		Fermented broth	—	>40 assay	

Prepolymers, see Fig. 1 and Table 1

3.2 Microbial Cells

Application of immobilized microbial cells are very interesting and important because microbial cells possess enzyme systems which catalyze multi-step reactions [3]. Microbial cells can be easily entrapped by prepolymer methods, and inactivation of the components of multi-enzyme systems may be avoided by using prepolymers as gel materials. Thus, different conditions of cells — dried, untreated, resting and growing — have been entrapped by prepolymer methods and applied to various reactions (Table 6).

Production of ATP by using the energy of glycolysis is very important for constructing bioreactors. Dried yeast cells entrapped by radical polymerization [27] or photo-crosslinking [38] of prepolymers were employed for the phosphorylation of adenosine with glucose as an energy source. Such ATP-generating systems could be combined with an ATP-requiring system in the same cells to produce cytidine diphosphate (CDP)-choline from cytidine monophosphate (CMP) and choline [39,40]. Entrapment of the cells with photo-crosslinkable resin prepolymers permitted the repeated use of the cells.

NADP was synthesized from NAD and ATP by the action of NAD kinase in *Brevibacterium ammoniagenes* cells entrapped by γ-irradiation [22]. The immobilized cells were stable over 30 batches of reaction (each reaction, 3 h).

Table 6. Microbial cells entrapped by prepolymer methods

Microorganism (condition)	Entrapment method (prepolymer)	Application	Ref.
Escherichia coli (untreated)	Urea linkage (urethane prepolymer)	Ammonium fumarate → L-Aspartic acid	20)
Escherichia coli (acetone-dried)	Photo-crosslinking (ENTE)	Assay of L-hydroxy amino acids	36, 37)
Baker's yeast (dried)	Radical polymerization (PEGM + acrylic acid + *N,N*-dimethylaminoethyl methacrylate)	Adenosine → ATP	27)
Baker's yeast (dried)	Photo-crosslinking (ENT)	Adenosine → ATP	38)
Brevibacterium ammoniagenes (freezed)	γ-Irradiation (PEGM + PVA)	NAD + ATP → NADP	22)
Hansenula jadinii (dried)	Photo-crosslinking (ENT)	Production of CDP-choline	39, 40)
Brevibacterium ammoniagenes (acetone-dried)	Photo-crosslinking (ENT)	Production of CoA	90)
Enterobacter aerogenes (thawed)	Photo-crosslinking (ENT and ENTP) and urea linkage (PU)	Uracil arabinoside + adenine → adenine arabinoside	41)
Escherichia coli (untreated)	Urea linkage (urethane prepolymers)	Penicillin G → 6-aminopenicillanic acid	19, 42)
Escherichia coli (untreated)	Polycondensation (epoxides)	Penicillin G → 6-aminopenicillanic acid	42)
Streptomyces clavuligerus (resting)	Crosslinking with dialdehyde (water-soluble PAAM having acylhydrazide group)	Production of cephalosporins	25)
Citrobacter freundii (acetone-dried)	Photo-crosslinking (ENTE)	Assay of cephalosporins	43)
Propionibacterium sp. (growing)	Photo-crosslinking (ENT) and urea linkage (PU)	Production of vitamin B_{12}	44)
Alcaligenes eutraphus (freeze-dried or untreated)	Urea linkage (PU)	Activity of hydrogenase	34)
Arthrobacter simplex (acetone-dried)	Photo-crosslinking (ENT and ENTP)	Hydrocortisone → prednisolone	9, 11)
Arthrobacter simplex (acetone-dried)	Urea linkage (PU)	Hydrocortisone → prednisolone	17, 18)
Nocardia rhodocrous (thawed)	Photo-crosslinking (ENTP etc.) and urea linkage (PU)	3β-Hydroxy-Δ^5-steroids → 3-keto-Δ^4-steroids	10, 45, 46)
Nocardia rhodocrous (thawed)	Urea linkage (PU)	Dehydrogenation of testosterone etc.	47, 48)
Nocardia rhodocrous (thawed)	Photo-crosslinking (ENT and ENTP)	4-Androstene-3,17-dione → 1,4-androstadiene-3,17-dione	11, 49)
Streptomyces roseochromogenes (untreated)	Photo-crosslinking (PEGM)	Dehydroepiandrosterone → 16α-hydroxydehydroepiandrosterone	6)
Curvularia lunata (living)	Photo-crosslinking (ENT)	Cortexolone → hydrocortisone	50, 51, 52)
Curvularia lunata (living) + *Arthrobacter simplex* (acetone-dried)	Photo-crosslinking (ENT and ENTP)	Cortexolone → prednisolone	53)

Rhizopus stolonifer (living)	Photo-crosslinking (ENT)	Progesterone → 11α-hydroxyprogesterone	54)
Corynebacterium sp. (living)	Photo-crosslinking (ENT and ENTP)	4-Androstene-3,17-dione → 9α-hydroxy-4-androstene-3,17-dione	55)
Rhodotorula minuta var. *texensis* (thawed)	Photo-crosslinking (ENT) and urea linkage (PU)	*dl*-Menthyl succinate → *l*-menthol + *d*-menthyl succinate	56)
Saccharomyces sp. (growing)	Photo-crosslinking (ENT etc.)	Production of ethanol	91)

ENT, hydrophilic photo-crosslinkable resin prepolymer; ENTE, emulsion-type photo-crosslinkable resin prepolymer; ENTP, hydrophobic photo-crosslinkable resin prepolymer; PAAM, polyacrylamide; PEGM, poly(ethylene glycol) dimethacrylate; PU, urethane prepolymer; PVA, poly(vinyl alcohol)

Several strains of *Escherichia coli*, especially hybrid strains whose penicillin acylase level was markedly enhanced by gene manipulation techniques, were entrapped with urethane prepolymers and epoxide resins, and the hydrolysis of penicillin G to 6-aminopenicillanic acid was examined [19,42]. Immobilized cells having markedly enhanced and stabilized penicillin acylase activity could be obtained by the combination of excellent mutants and suitable gels. *De novo* synthesis of cephalosporins was tried with *Streptomyces clavuligerus* cells entrapped with linear, water-soluble polyacrylamide [25]. The cells entrapped with the prepolymer showed much higher activity than the cells entrapped with acrylamide monomer.

A very complicated compound, vitamin B_{12}, can also be synthesized *de novo* by entrapped growing cells of *Propionibacterium* sp. [44]. A hydrophilic photo-crosslinkable resin prepolymer (ENT) and a urethane prepolymer (PU) could be applied to the entrapment of living bacterial cells. *Alcaligenes eutrophus* cells entrapped by PU were found to show the activity of hydrogenase [34].

Microbial cell tubes prepared with the emulsion-type photo-crosslinkable resin prepolymer (ENTE-1) were employed in continuous flow systems to measure cephalosporins [43] and *L*-hydroxy amino acids such as *L*-threonine and *L*-serine [36,37].

Immobilization in suitable gels renders microbial cells significantly tolerant against unfavourable effects of organic solvents. Applications of such immobilized microbial cells to the bioconversion of hydrophobic compounds and water-insoluble compounds are very important from a practical point of view. In these cases, introduction of organic solvents into reaction systems is essential to dissolve substrates and/or products, and to construct homogeneous reaction systems. Various reactions have been successfully carried out by using microbial cells entrapped with prepolymers in water-organic cosolvent systems (production of adenine arabinoside [41], dehydrogenation of a steroid [9,11,17,18] and hydroxylation of steroids [6,50–55]) and in organic solvent systems (dehydrogenation of steroids [10,11,45–49] and optical resolution of *dl*-menthol [56]). Living and treated cells could be used for these purposes. Details will be described later.

3.3 Cellular Organelles

Cellular organelles have their own distinctly organized functions for metabolizing different cellular substances. This means that these organelles will become excellent biocatalysts for carrying out the well-organized multi-step reactions, if they are isolated intact and immobilized without the loss of their native characteristics [57,58]. Prepolymer methods have been proved to be suitable for the entrapment of organelles because the entrapment can be achieved under very mild conditions. Although only a few reports have been published, the results are summarized in Table 7.

Thylakoids (chloroplast membranes) isolated from lettuce were immobilized by various methods, and their activity of photochemical oxygen evolution was examined. Although high activity was obtained with the organelles entrapped in albumin crosslinked with glutaraldehyde, those entrapped with a photo-crosslinkable resin prepolymer (ENT) and a urethane prepolymer (PU) showed moderate activity [60]. Spinach chloroplasts were entrapped by γ-irradiation with different monomers and prepolymers [23]. Poly(ethylene glycol) diacrylate (PEGA) and poly(ethylene glycol)

Table 7. Cellular organelles entrapped by prepolymer methods

Organelle (source)	Entrapment method (prepolymer)	Application	Ref.
Chromatophores (*Rhodopseudomonas capsulata*)	Urea linkage (PU) and photo-crosslinking (ENT)	ATP formation by photo-phosphorylation	59)
Thylakoids (lettuce)	Urea linkage (PU) and photo-crosslinking (ENT)	Photolysis of water	60)
Chloroplasts (spinach)	γ-Irradiation (PEGA and PEGM)	Photolysis of water	23)
Mitochondria (acetate-grown *Candida tropicalis*)	Photo-crosslinking (ENT) and urea linkage (PU)	Activity of adenylate kinase	61)
Peroxisomes or microbodies (methanol-grown *Kloeckera* sp.)	Photo-crosslinking (ENT)	Fundamental study	7,62)
	Photo-crosslinking (ENT)	Activities of alcohol oxidase, catalase and *D*-amino acid oxidase	12,62,63)
	Urea linkage (PU)	Activities of alcohol oxidase and catalase	18)

ENT, hydrophilic photo-crosslinkable resin prepolymer; PEGA, poly(ethylene glycol) diacrylate; PEGM, poly(ethylene glycol) dimethacrylate; PU, urethane prepolymer

dimethacrylate (PEGM) were found to be suitable gel materials. Both thylakoids and chloroplasts were stabilized by entrapment.

Yeast mitochondria were entrapped with a photo-crosslinkable resin prepolymer (ENT) and a urethane prepolymer (PU). Although the entrapped organelles lost the activity of oxidative phosphorylation, these preparations exhibited a stabilized activity of adenylate kinase [61].

Peroxisomes (microbodies) from methanol-grown yeast cells were also entrapped successfully with a photo-crosslinkable resin prepolymer (ENT) [7,63] and a urethane prepolymer (PU) [18]. The entrapped organelles showed activities of alcohol oxidase, catalase and D-amino acid oxidase, the results suggesting the possible application of entrapped peroxisomes to the analyses or degradation of alcohols, hydrogen peroxide and D-amino acids [63]. Synergic action of alcohol oxidase and catalase in the oxidation of methanol to formaldehyde has also been proved by using the immobilized organelles [7].

Bacterial chromatophores, which are not true organelles but an important biological photosystem, were entrapped by various methods. In addition to the crosslinking of albumin with glutaraldehyde, entrapment of chromatophores with a urethane prepolymer (PU) yielded a high activity of photophosphorylation [59]. The ATP-synthesizing activity was also stabilized by entrapment.

4 Effects of the Physico-Chemical Properties of Prepolymers

One of the advantages of prepolymer methods is the easy preparation of gels with different properties by selecting suitable prepolymers which were synthesized chemically under rather drastic conditions in the absence of biocatalysts.

4.1 Chain-length

The chain-lengths of prepolymers can be changed by selecting poly(ethylene glycol) (in the case of PEGM and ENT), poly(propylene glycol) (in the case of ENTP) or polyether diol (in the case of PU) of different molecular weight (see, Figs. 1 and 2, and Tables 1 and 3). For example, the chain-length of a prepolymer containing poly(ethylene glycol)-1000 (average Mw, about 1,000) is about 10 nm, and that containing poly(ethylene glycol)-2000 (average Mw, about 2,000) is about 20 nm. In general, a loose net-work of gels prepared from longer chain prepolymers renders diffusion of substrates and products favourable. However, leakage of enzyme proteins, especially those with small molecular weight, sometimes becomes a significant problem with the gels which have a loose net-work [13].

When invertase was entrapped with PEGM-1000 to PEGM-4000 [4] or PU of Mw 2,000 to 5,000 [16], the apparent enzyme activity was highest in the case of the enzyme entrapped with the longest-chain prepolymer. A similar result was also obtained when catalase was entrapped with ENT-1000 to ENT-6000 [8].

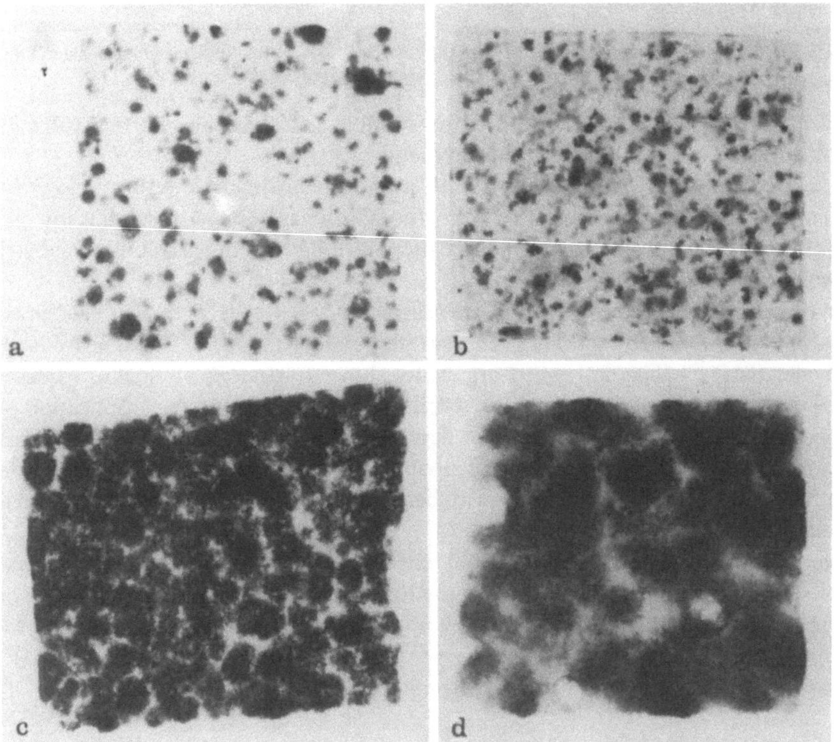

Fig. 3a—d. Development of *Curvularia lunata* mycelia inside photo-crosslinked resin gels prepared from prepolymers of different chain-length [52]. Entrapped spores were cultivated for 60 h. **a**, ENT-1000-entrapped mycelia; **b**, ENT-2000-entrapped mycelia; **c**, ENT-4000-entrapped mycelia; **d**, ENT-6000-entrapped mycelia. Size of each gel, 5 × 5 mm

Interesting phenomena were observed when fungal spores were entrapped, allowed to germinate and grow inside gel matrices[52,54]. Development of mycelia was abundant in gels formed from prepolymers of longer chain-length, while the growth was inhibited in gels of tight net-work (Fig. 3). Leakage of mycelia was often observed with gels of a loose net-work. In accordance with these facts, high and stable activities of steroid hydroxylation were obtained with mycelia entrapped with medium chain prepolymers.

As described above, the size of gel net-work is sometimes an important factor affecting the activity and stability of entrapped biocatalysts. Prepolymer methods offer an easy selection of adequate gels depending on the kind of biocatalysts to be used.

4.2 Ionic Property

Anionic or cationic property of gels has a significant influence on the apparent activities of gel-entrapped biocatalysts through an electrostatic interaction with the charged substrate. Furthermore, the optimum reaction pH may be shifted by using gels with different ionic properties.

When choline oxidase was entrapped with photo-crosslinkable resin prepolymers and used as an enzyme electrode for the assay of choline, the enzyme entrapped with an anionic prepolymer exhibited significantly high activity. On the other hand, the enzyme immobilized in cationic gels showed markedly low activity [13]. The results can be explained by the different affinity of the substrate choline with cationic property to the gels with different ionic properties. The activity of uricase was rather reduced in gels with anionic properties.

Shift of the optimum reaction pH was also observed with the photo-crosslinked gels of different ionic character [41].

4.3 Hydrophilicity-Hydrophobicity Balance

In general, bioreactions with hydrophilic substrates are carried out in aqueous solutions. Hydrophilic properties of gels are favourable for such reaction systems. On the other hand, organic solvent systems are suitable for bioconversions of hydrophobic compounds. In such cases, hydrophobic gels would favour the transformation of highly hydrophobic substances due to high affinity of the compounds for the gels. One of the most representative examples is the successful conversion of 3β-hydroxy-Δ^5-steroids to the corresponding 3-keto-Δ^4-steroids in a non-polar organic solvent systems [10,46,48,64], indicating the advantage of hydrophobic gels, as described in detail later. A similar effect of gel hydrophobicity was also observed in the optical resolution of *dl*-menthol [56], ester exchange of triglyceride [31] and hydrolysis of triglyceride [33]. It was found that the activities of the entrapped biocatalysts corresponded closely to the partitions of the substrates between the gels and the external solvents [10].

Prepolymer methods, especially those with photo-crosslinkable resin prepolymers and urethane prepolymers, again offer easy selection of gels which have an optional hydrophilicity-hydrophobicity balance by using different types of prepolymers or by mixing these prepolymers.

5 Bioconversions in the Presence of Organic Solvents

Enzymatic reactions, even in cases where water-insoluble, lipophilic compounds were substrates, used to be carried out in aqueous systems, because biocatalysts were believed to be unstable in organic solvents. However, it will be desirable to perform enzymatic reactions with lipophilic compounds in mixtures of water and suitable organic cosolvents, or in appropriate organic solvent systems, provided the catalytic activities of the biocatalysts are maintained under such reaction conditions. The use of organic solvents will improve the poor aqueous solubility of substrates or other reaction components of hydrophobic nature, and shift an unfavourable thermodynamic equilibrium to a desired direction if necessary [65,66]. Immobilization often considerably enhances the stability of biocatalysts against denaturation by organic solvents [67,68]. Thus, bioconversions of water-insoluble compounds, such as steroids, have been achieved by immobilized biocatalysts. For the conversion of lipophilic compounds, several solvent systems are applicable [69,70]: water/water-miscible organic solvent homogeneous systems, water/water-immiscible organic solvent two-phase systems and organic solvent systems. However, two-phase systems are inconvenient for carrying out the reactions continuously.

5.1 Water-Organic Cosolvent Systems

Water/water-miscible organic solvent systems have been widely employed to dissolve water-insoluble, lipophilic compounds to prepare homogeneous reaction systems and to shift reaction equilibrium to a desired direction, especially to the synthetic direction with hydrolyzing enzymes. For the bioconversion of steroids, relatively low concentrations of organic cosolvents such as methanol [71,72], ethanol [73] and *N,N*-dimethylformamide [74] were used to dissolve substrates and/or products. To direct proteases to the synthesis of peptide bonds, high concentrations of organic solvents were used [75,76]. Such reactions have been applied to the plastein synthesis [77,78], to the peptidation reaction for yielding human insulin from porcine insulin [79-81],

Table 8. Bioconversions in water-organic cosolvent systems by microbial cells entrapped with prepolymers

Microorganism (condition)	Organic cosolvent	Application	Ref.
Arthrobacter simplex (acetone-dried)	10% Methanol	Δ^1-Dehydrogenation of hydrocortisone	9, 11, 17
Streptomyces roseochromogenes (untreated)	0.25% *N,N*-Dimethylformamide	16α-Hydroxylation of dehydroepiandrosterone	6
Curvularia lunata (living)	2.5% Dimethyl sulfoxide	11β-Hydroxylation of cortexolone	50-53
Rhizopus stolonifer (living)	2.5% Dimethyl sulfoxide	11α-Hydroxylation of progesterone	54
Corynebacterium sp. (living)	15% Dimethyl sulfoxide	9α-Hydroxylation of 4-androstene-3,17-dione	55
Enterobacter aerogenes (thawed)	40% Dimethyl sulfoxide	Synthesis of adenine arabinoside	41

and so on. Biosyntheses of α-hydroxynitriles [82] and urea [83] have also been facilitated in the presence of organic cosolvents.

Microbial cells entrapped with various prepolymers have also been applied to the bioconversions of lipophilic or water-insoluble compounds in water-organic cosolvent systems (Table 8).

Acetone-dried cells of *Arthrobacter simplex*, entrapped with photo-crosslinkable resin prepolymers [9] or urethane prepolymers [17], catalyzed Δ^1-dehydrogenation of hydrocortisone to form prednisolone. Introduction of organic cosolvents (10%), such as methanol, ethylene glycol, propylene glycol, trimethylene glycol and glycerol, stimulated significantly the reaction of the entrapped cells together with solubilizing capacity of substrate and product. The entrapment markedly enhanced the stability of the cells at high concentrations of the organic cosolvents and maintained this stability during long-term operation.

Hydroxylation of steroids by entrapped resting or living cells was also carried out in the presence of organic cosolvents (Fig. 4)[6,50–55]. Cortexolone (Reichstein's Compound S) was hydroxylated at the 11β-position by the photo-crosslinked gel-entrapped mycelia of *Curvularia lunata*, which were derived from the spores germinated and developed inside gel matrices [50–52]. Dimethyl sulfoxide or methanol

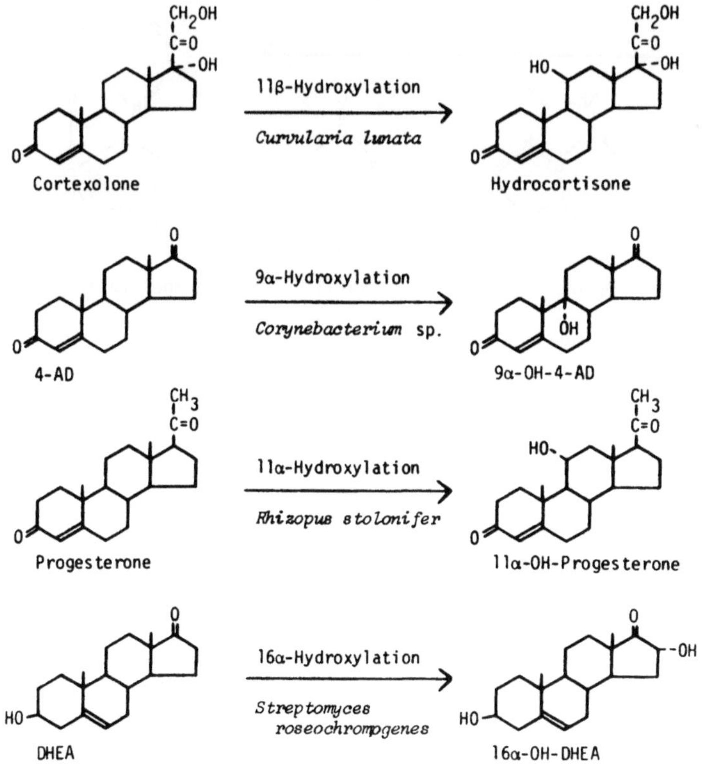

Fig. 4. Hydroxylation of steroids by immobilized microbial cells in water-organic cosolvent systems. Abbreviations used: 4-AD, 4-androstene-3,17-dione; DHEA, dehydroepiandrosterone

at 2.5% was insufficient to dissolve the substrate (cortexolone) but was sufficient to dissolve the product (hydrocortisone). The hydroxylation system in the entrapped mycelia could be reactivated by incubating the mycelium-entrapping gels in a nutrient medium containing cortexolone as an inducer. The entrapped mycelia were far more stable than the free counterparts, and could be utilized repeatedly for at least 50 batches of the reaction (total operational period; 100 days). This reaction system could be connected sequentially with the Δ^1-dehydrogenation system of the *A. simplex* cells to produce prednisolone from cortexolone [53]. Spores of *Rhizopus stolonifer* were also entrapped with photo-crosslinkable resin prepolymers, and allowed to germinate and develop inside gel matrices [54]. The entrapped fungal mycelia thus obtained hydroxylated progesterone at the 11α-position to form 11α-hydroxyprogesterone. In this case also, the enzyme system in the entrapped mycelia was reactivated by incubation in a nutrient medium. Entrapped living cells of *Corynebacterium* sp. were more tolerant to organic solvents than the entrapped fungal mycelia. Thus, 9α-hydroxylation of 4-androstene-3,17-dione to yield 9α-hydroxy-4-androstene-3,17-dione was performed with dimethyl sulfoxide at 15% in a nutrient medium [55]. The reaction in a buffer solution instead of the medium resulted in a rapid decrease in the hydroxylation activity.

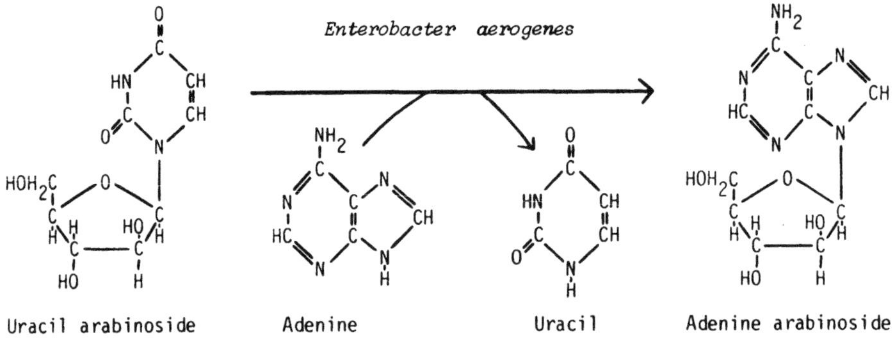

Fig. 5. Synthesis of adenine arabinoside by *Enterobacter aerogenes*

The synthesis of adenine arabinoside, an anti-viral antibiotic, from uracil arabinoside and adenine (Fig. 5) was achieved by the entrapped cells of *Enterobacter aerogenes* in a system containing 40% of dimethyl sulfoxide [41]. Although the reaction could be carried out successfully in an aqueous system [84], the productivity was low because of the low solubility of adenine and adenine arabinoside in water. By selecting an organic cosolvent based on the criteria of stability of the enzyme system and solubility of adenine arabinoside, dimethyl sulfoxide at 40% was employed as an organic cosolvent. Dosage of a high concentration of the substrates to the optimized reaction system resulted in the formation of a high concentration of the product, which was easily recovered from the reaction solution only by cooling and filtering. Entrapment of *E. aerogenes* cells with photo-crosslinkable resin prepolymers or urethane prepolymers remarkably enhanced the operational stability of the enzyme system during repeated reactions. Thus, the entrapped cells could be used for at least 35 days in the presence of 40% dimethyl sulfoxide at 60 °C without any loss of the enzyme activity (Fig. 6).

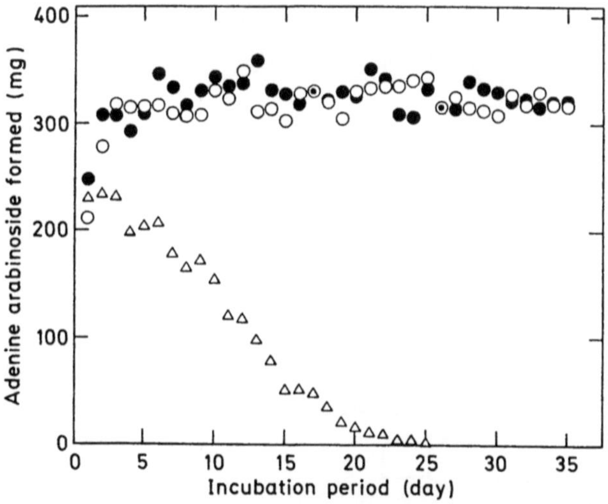

Fig. 6. Repeated use of *Enterobacter aerogenes* cells for production of adenine arabinoside[41]. Each reaction was carried out for 24 h at 60 °C in the presence of 40% dimethyl sulfoxide. (○), ENT-4000-entrapped cells; (●), PU-6-entrapped cells; (△), free cells

As described above, water/organic cosolvent systems could be applied to the bioconversions which are catalyzed by immobilized biocatalysts. Immobilization often gives markedly enhanced stability to biocatalysts. This effect is probably ascribable to maintenance of the active conformation of enzyme molecules by rigid interaction between biocatalysts and supports which prevents inactivation by organic solvents, and to effective coating of microbial cells which protects leakage of enzymes from gel-entrapped cells.

5.2 Organic Solvent Systems

Although water-immiscible organic solvents have been employed for bioreactions to increase the solubility of substrates and products of a hydrophobic nature, most cases deal with water-organic solvent two-phase systems[70]. Only a limited numbers of papers have described the single use of organic solvents in bioreactions.

Klibanov et al. [85] reported pioneering work on the use of organic solvents in the synthesis of *N*-acetyl-L-tryptophan ethyl ester from *N*-acetyl-*L*-tryptophan and ethanol by chymotrypsin covalently bound to porous glass. In this case, chloroform, benzene or ether was used as the reaction solvent. Toluene or carbon tetrachloride was also employed in the dehydrogenation of cholesterol to cholestenone by DEAE-cellulose-adsorbed cells of *Nocardia erythropolis* [86].

The present authors have extensively investigated bioconversions of lipophilic compounds by immobilized biocatalysts in organic solvent systems (Table 9). The activities of the immobilized biocatalysts were found to be affected by the hydrophilicity-hydrophobicity balance of the gels, the hydrophobicity of substrates and the polarity of the reaction solvents [46, 48, 64].

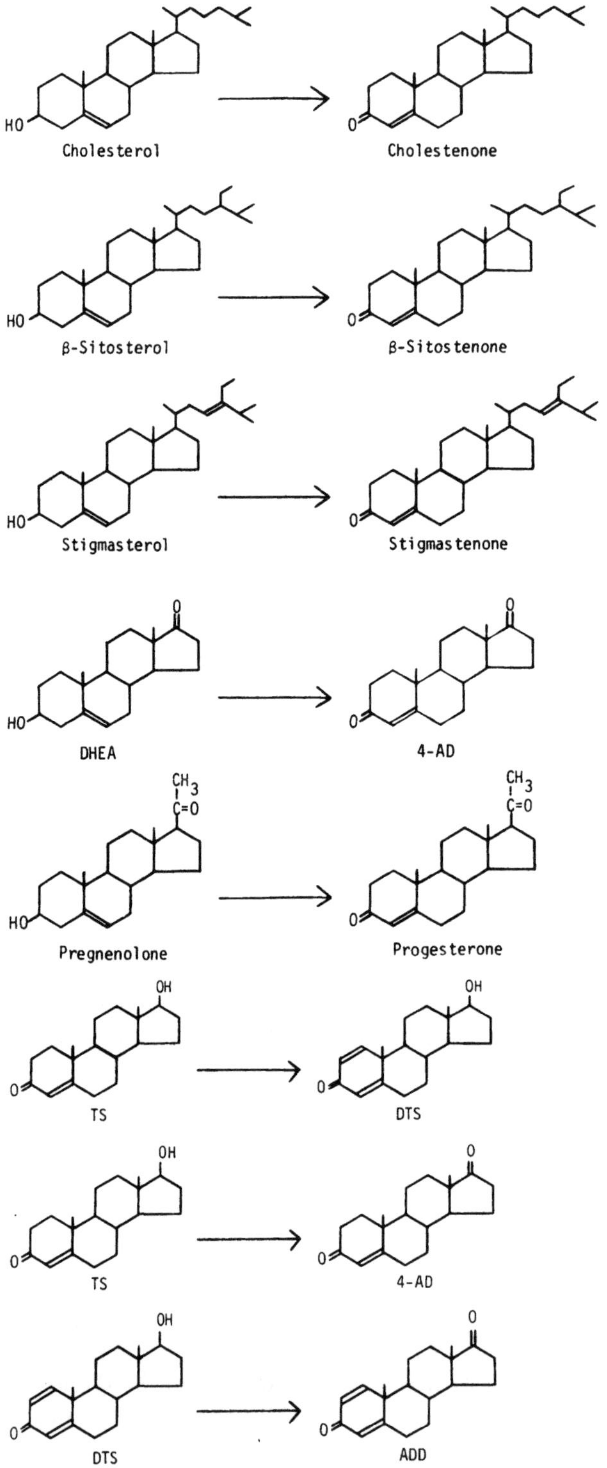
HO
O
Cholesterol
Cholestenone
β-Sitosterol
β-Sitostenone
Stigmasterol
Stigmastenone
DHEA
4-AD
CH3
C=O
Pregnenolone
Progesterone
OH
TS
DTS
TS
4-AD
DTS
ADD

Fig. 7.

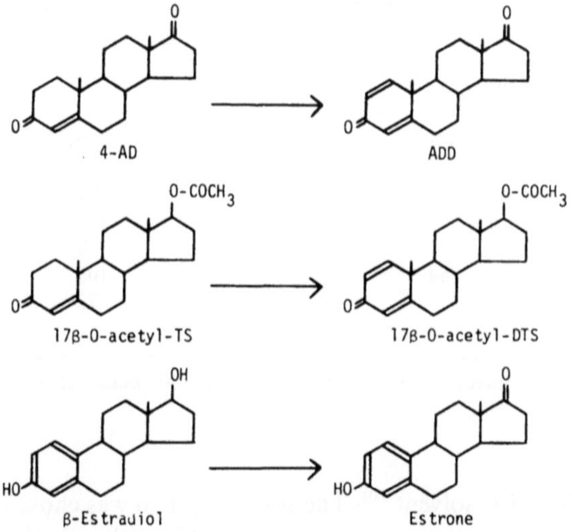

Fig. 7. Steroid conversions mediated by *Nocardia rhodocrous* in organic solvent systems [47,48]. Abbreviations used: ADD, 1,4-androstadiene-3,17-dione; 4-AD, 4-androstene-3,17-dione; DHEA, dehydroepiandrosterone; DTS, Δ^1-dehydrotestosterone; TS, testosterone

As shown in Fig. 7, *Nocardia rhodocrous* cells mediate different types of bioconversions of a variety of steroids; Δ^1-dehydrogenation, 3β-hydroxysteroid dehydrogenation and 17β-hydroxysteroid dehydrogenation [10,11,45–49]. In order to study the effect of gel hydrophobicity, — that is, the influence of affinity between hydrophobic substrates and gels entrapping biocatalysts, a very simple parameter, the partition coefficient (a ratio of substrate concentration between the gels and the external solvent) was employed.

The effect of gel hydrophobicity was mainly investigated in the conversion of 3β-hydroxy-Δ^5-steroids into 3-keto-Δ^4-steroids in a water-saturated mixture of

Table 9. Bioconversions in organic solvent systems by biocatalysts entrapped with prepolymers

Biocatalyst	Organic solvent	Application	Ref.
Nocardia rhodocrous cells	Benzene-*n*-Heptane (1:1 by volume)	Δ^1-Dehydrogenation of 4-androstene-3,17-dione	11,49
	Benzene-*n*-Heptane (4:1 by volume)	Δ^1-Dehydrogenation of testosterone	47
	Benzene-*n*-Heptane (1:1 by volume)	3β-Hydroxysteroid dehydrogenation	10
	Benzene-*n*-Heptane (4:1 by volume)	17β-Hydroxysteroid dehydrogenation	47
	Chloroform-*n*-Heptane (1:1 by volume)	3β-Hydroxysteroid dehydrogenation	45
Rhodotorula minuta cells	*n*-Heptane	Resolution of *dl*-menthol	56
Rhizopus delemar lipase	*n*-Hexane	Interesterification of triglyceride	31,32

Table 10. Effect of gels on steroid transformation by entrapped cells of *Nocardia rhodocrous* in non-polar solvent [10, 48]

Gel	Relative activity (%) on			
	Cholesterol	β-Sitosterol	Stigmasterol	DHEA
None	100	100	100	100
ENTP-2000 (hydrophobic)	102	83	111	107
ENT-4000 (hydrophilic)	0	0	0	78
PU-3 (hydrophobic)	77	54	85	105
PU-6 (hydrophilic)	0	—	0	72
PU-9 (hydrophilic)	0	0	—	—

DHEA, dehydroepiandrosterone; Reaction solvent, water-saturated benzene-*n*-heptane (1:1 by volume)

benzene and *n*-heptane (1:1 by volume) as solvent [10]. The solvent system was chosen on the basis of the following criteria: substrates and products are sufficiently soluble; the enzyme system is not damaged; and the solvent does not cause the gels to swell. As shown in Table 10, *N. rhodocrous* cells entrapped in hydrophobic gels, such as ENTP-2000 and PU-3, converted dehydroepiandrosterone (DHEA) into 4-androstene-3,17-dione (4-AD) with the high reaction rate comparable to that of the free cells. On the other hand, the cells entrapped in hydrophilic gels, ENT-4000 and PU-6, were less active. Fig. 8b shows the relationship between the relative activity of entrapped cells and the partition coefficient of DHEA, both of which changed depending on the hydrophobicity of gels. The abscissa shows the mixing ratio of the

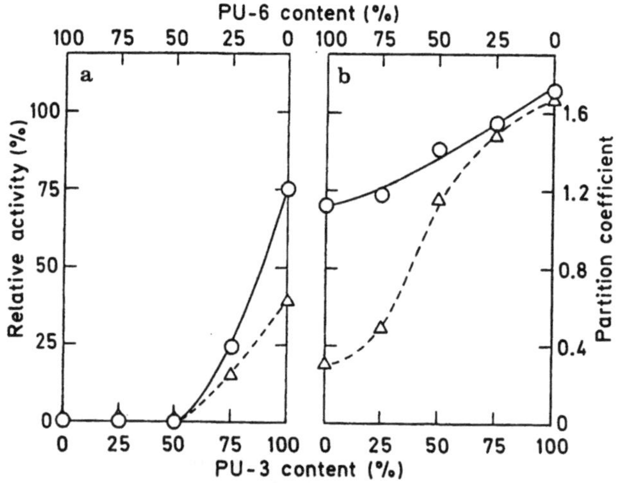

Fig. 8a and b. Effect of gel hydrophobicity on steroid conversion activity of gel-entrapped *Nocardia rhodocrous* cells and on the partition coefficient of the substrate [10,48]. **a**, cholesterol; **b**, dehydroepiandrosterone. The reaction was carried out in water-saturated benzene-*n*-heptane (1:1 by volume). (○), Relative activity to that of the free cells; (△), partition coefficient

hydrophobic prepolymer PU-3 and the hydrophilic prepolymer PU-6. A close relationship between the relative activity of the gel-entrapped cells and the partition coefficient of the substrate can be observed. In the cases of the transformation of cholesterol, β-sitosterol and stigmasterol (substrates that are more lipophilic than DHEA) to the corresponding 3-keto-Δ^4-steroids, a much more clear-cut interrelationship was observed between gel hydrophobicity and the activity of the gel-entrapped cells (Table 10). Only the cells entrapped with hydrophobic prepolymers (ENTP-2000 and PU-3) exhibited the catalytic activity. This phenomenon was also confirmed when the hydrophobicity of the gels was changed by mixing PU-3 and PU-6 (Fig. 8a). No activity of cholesterol conversion, in accordance with the low partition coefficient of cholesterol, was observed at the low ratio of PU-3, a hydrophobic prepolymer.

In addition to gel hydrophobicity and substrate hydrophobicity, the polarity of the reaction solvents also produced a marked effect on the conversion of steroids [45]. *N. rhodocrous* cells entrapped in hydrophilic gels could not transform cholesterol in a non-polar solvent, e.g. benzene-*n*-heptane (1:1 by volume). However, if benzene is replaced by chloroform, the hydrophilic gel-entrapped cells become active, in accordance with the increased partition coefficient (Fig. 9). However, enhancement of

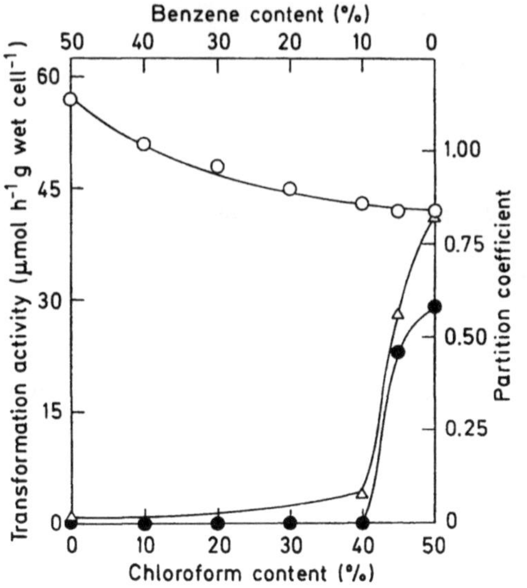

Fig. 9. Effect of solvent polarity on cholesterol transformation by hydrophilic gel-entrapped *Nocardia rhodocrous* cells [45]. (○), Activity of free cells; (●), activity of ENT-4000-entrapped cells; (△), partition coefficient of cholesterol. Solvent, *n*-heptane (50%)-benzene-chloroform

solvent polarity fairly lowered the activity and stability of the free cells and, subsequently, those of the entrapped cells. In the water-saturated mixture of chloroform and *n*-heptane (1:1 by volume), little effect of gel hydrophobicity was observed in the conversion of cholesterol and DHEA, differing from the results obtained in the non-polar solvent system shown in Fig. 8. The hydrophobic gel-entrapped cells had rather low activity in the conversion of pregnenolone, a less hydrophobic substrate which was not soluble in benzene-*n*-heptane. In spite of these facts, the usefulness of hydrophobic gels is clear in the transformation of highly lipophilic compounds because the transformation activity of the gel-entrapped cells is usually high and stable in less polar solvents. In conclusion, it would be very important for successful bioconversions of lipophilic compounds to select gel materials with suitable hydrophobicity and reaction solvents with proper polarities, both of which should be determined according to substrate hydrophobicity. For the entrapment of biocatalysts, the prepolymer methods are very useful because these methods offer very easy selection of gel hydrophobicity.

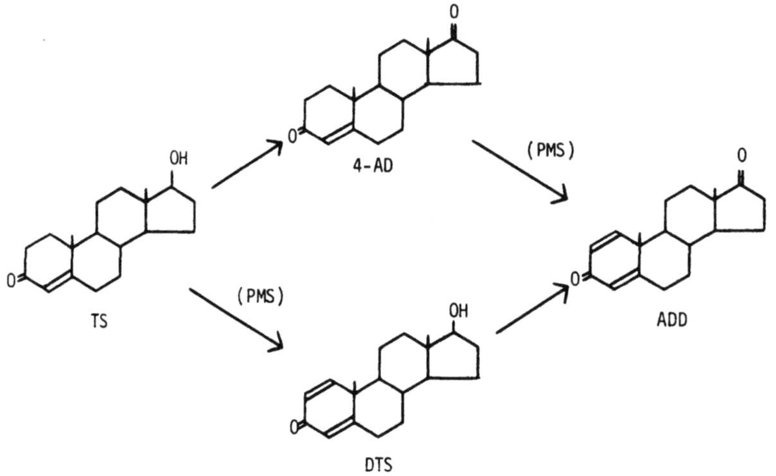

Fig. 10. Conversion of testosterone catalyzed by *Nocardia rhodocrous* in organic solvent system [47]. Abbreviations used: ADD, 1,4-androstadiene-3,17-dione; 4-AD, 4-androstene-3,17-dione; DTS, Δ^1-dehydrotestosterone; TS, testosterone; PMS, phenazine methosulfate

Gel hydrophobicity also affected the conversion routes from testosterone (TS) to 1,4-androstadiene-3,17-dione (ADD) [47]. Figure 10 illustrates the transformation pathway of TS into ADD mediated by *N. rhodocrous* in water-saturated benzene-*n*-heptane (4:1 by volume). In the presence of an electron acceptor, such as phenazine methosulfate (PMS), the free bacterial cells converted TS to ADD via two diverse routes. As shown in Fig. 10, the 17β-dehydrogenation product, 4-androstene-3,17-dione (4-AD), and the Δ^1-dehydrogenation product, Δ^1-dehydrotestosterone (DTS), appeared as the intermediates in ADD formation. In this reaction, the Δ^1-dehydrogenation absolutely required PMS, whereas the 17β-dehydrogenation could proceed without the exogenous electron acceptor, although PMS stimulated the reaction. When the cells were entrapped in gels of different hydrophilicity or hydrophobicity, the properties of the gels produced striking effects on the conversion routes. With hydrophobic (PU-3-rich) gel-entrapped cells, 4-AD was formed as a major reaction product. On the other hand, DTS was the main product with hydrophilic (PU-6-rich) gel-entrapped cells. This different profile in dehydrogenation products can be explained by a marked difference in the affinity of PMS, a hydrophilic compound, to the hydrophobic gels and the hydrophilic ones. With hydrophilic gel-entrapped cells, the Δ^1-dehydrogenation of TS to yield DTS is stimulated by PMS absorbed inside the gels and DTS so accumulated inhibits 17β-hydroxysteroid dehydrogenase from converting DTS to ADD even at a low concentration. On the other hand, PMS was hardly uptaken into the cells entrapped within the hydrophobic gels and hence the 17β-dehydrogenation of TS becomes the main route.

These results indicate that the dehydrogenation reaction at two distinct positions of TS — Δ^1-dehydrogenation and 17β-hydroxy group dehydrogenation — can be controlled by selecting the hydrophilic or hydrophobic nature of gels entrapping *N. rhodocrous* cells. The selective formation of a desired product among diverse products from a single substrate by the appropriate use of hydrophilic or hydrophobic gels to entrap cells would be applicable to the bioconversions of many organic compounds.

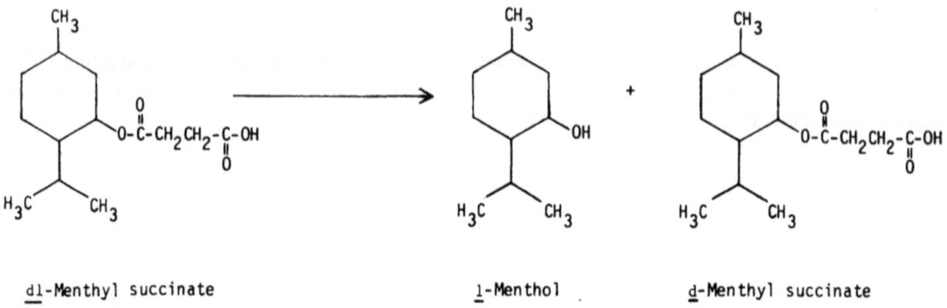

Fig. 11. Stereoselective hydrolysis of *dl*-menthyl succinate catalyzed by *Rhodotorula minuta* var. *texensis*

The stereoselective hydrolysis of *dl*-menthyl succinate by gel-entrapped cells of *Rhodotorula minuta* var. *texensis* (Fig. 11) was carried out successfully in water-saturated *n*-heptane [56]. *l*-Menthol, the compound having a peppermint flavour and useful in the food and pharmaceutical industries, can be obtained by stereospecific hydrolysis of an appropriate ester of chemically synthesized *dl*-menthol. The ammonium salt of *dl*-menthyl succinate is water-soluble, and its stereoselective hydrolysis could be achieved in an aqueous system by using free cells of the yeast which possesses esterase activity. However, due to poor solubility in aqueous buffers, *l*-menthol formed accumulated on the surface of the yeast cells, thus decreasing the activity. To prevent the accumulation of *l*-menthol on the cell surface, various kinds of water-miscible organic solvents were tested as cosolvents. However, the hydrolytic activity of the yeast cells was reduced in the presence of organic cosolvents. After many attempts to use free cells in organic solvents and to find an appropriate two-phase system consisting of a potassium phosphate buffer and an organic solvent, the combination of gel-entrapped cells and water-saturated *n*-heptane was finally employed to obtain a homogeneous reaction system. Although the effect of gel hydrophobicity was not so remarkable as in the case of steroid conversion, the activity of the gel-entrapped cells increased along with increased gel hydrophobicity. Fig. 12 shows the comparison of operational stability between free cells and

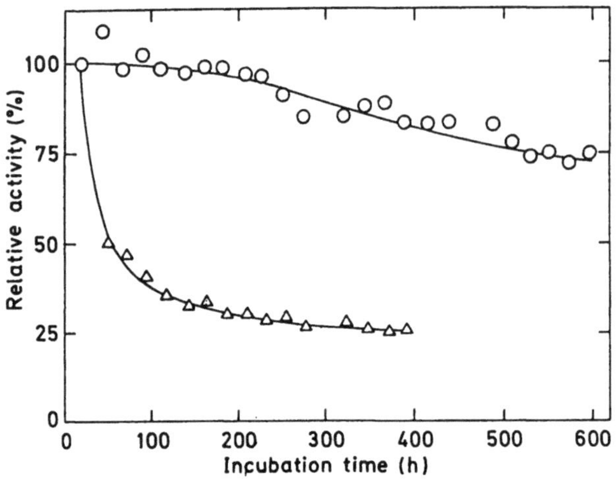

Fig. 12. Repeated use of free and entrapped *Rhodotorula minuta* cells on *l*-menthol production [48,56] Each reaction was carried out for 23 h in water-saturated *n*-heptane (entrapped cells) or in water-*n*-heptane (free cells). (○), PU-3-entrapped cells, (△), free cells

PU-3-entrapped cells over repeated reactions. The half-life of the free cells was 2 days, while that of the entrapped cells was estimated to be 63 days. Thus, immobilization greatly improved the operational stability of the hydrolytic enzyme in the yeast cells. The optical purity of the product was also constantly maintained at 100% even after a long-term operation. Figure 13 illustrates schematically the large-scale production of *l*-menthol by the use of immobilized *R. minuta* cells. The starting material, *dl*-mentol, was converted into the succinate ester, then hydrolyzed stereospecifically. The yield was 86%. *d*-Menthyl succinate remained and succinic acid liberated were recycled as shown in the figure.

The next examplė is the interesterification of triacylglyceride by lipase. The reformation of olive oil to cacao butter-like fat has been tried by exchanging oleic

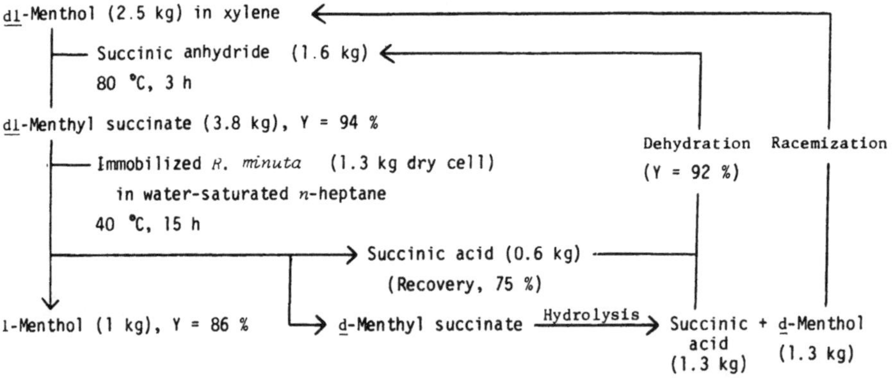

Fig. 13. Large scale production of *l*-menthol by immobilized *Rhodotorula minuta* cells [56].

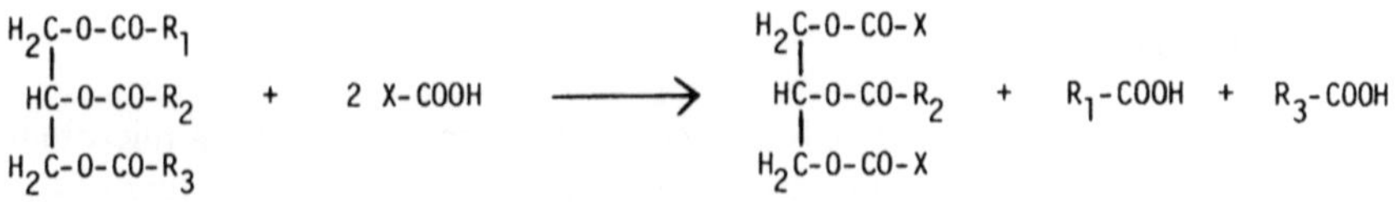

Fig. 14. Ester exchange reaction of triacylglyceride mediated by *Rhizopus delemar* lipase

acid moieties in positions 1 and 3 of the triglyceride with saturated fatty acids, such as stearic acid, by 1 and 3 position-specific lipase from *Rhizopus delemar* (Fig. 14) [87]. In this reaction, it is very important to control the water content in the reaction system, because while the ester exchange reaction cannot be initiated without water, hydrolysis of ester is preferential when the concentration of water is high. Therefore, water-saturated *n*-hexane was selected as the reaction solvent, taking into consideration the activity and stability of the enzyme as well as the solubility of the reactants. To provide water in the vicinity of the enzyme, lipase was first adsorbed on a suitable porous support, such as Celite, with a controlled amount of water. When the Celite-adsorbed lipase was entrapped by different prepolymers [31], the enzyme entrapped by a hydrophobic photo-crosslinkable resin prepolymer, ENTP-2000, showed the highest activity of interesterification, about 75 % of that of lipase when simply adsorbed onto Celite. On the other hand, the hydrophilic gel-entrapped enzyme exhibited only low activity. This result again indicates the usefulness of hydrophobic gels in the bioconversion of lipophilic compounds in a non-polar solvent. Entrapment markedly enhanced the operational stability of lipase, the enzyme losing only 10% of the original activity after 12 reaction batches (operational period, 12 days) (Fig. 15).

The results mentioned above, together with immobilized chymotrypsin reported by Klibanov et al.[85], demonstrate that not only microbial cells but also enzymes can be employed as biocatalysts in organic solvent systems after appropriate immobilization.

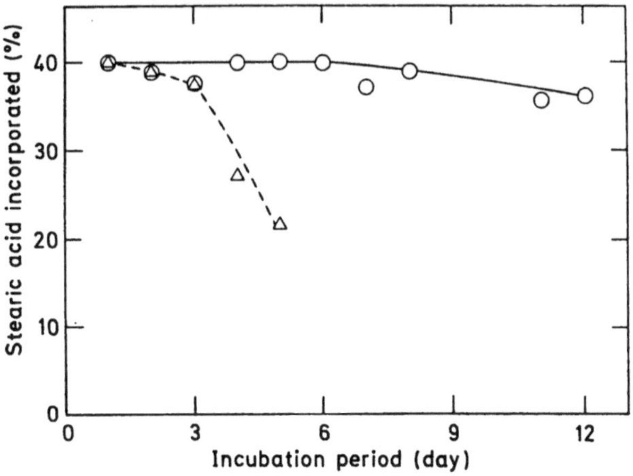

Fig. 15. Repeated use of lipase preparations for interesterification of triglyceride. [31,32]. Each reaction was carried out at 40 °C for 24 h in water-saturated *n*-hexane. (△), Celite-adsorbed lipase; (○), ENTP-2000-entrapped Celite-adsorbed lipase

6 Entrapped Living Cells

In addition to immobilized treated cells and enzymes, immobilized living microbial cells, especially immobilized growing cells, have aroused great interest because of the superior stability that results from the self-regenerating nature of catalytic systems. Immobilized growing cells can catalyze not only single conversions and salvage syntheses, but also syntheses of various complex compounds which are produced at present by conventional bioprocess techniques.

Various types of supports have been employed for preparing immobilized living microbial cells which are then used to achieve different conversions and biosynthetic reactions [3, 88]. Photo-crosslinkable resin prepolymers and urethane prepolymers are also good gel materials for entrapping living cells which can then propagate inside the gel matrices. Fig. 16 shows typical examples of a bacterium (*Corynebacterium* sp.), a yeast (*Candida tropicalis*) and a fungus (*Curvularia lunata*) grown in polyurethane gels.

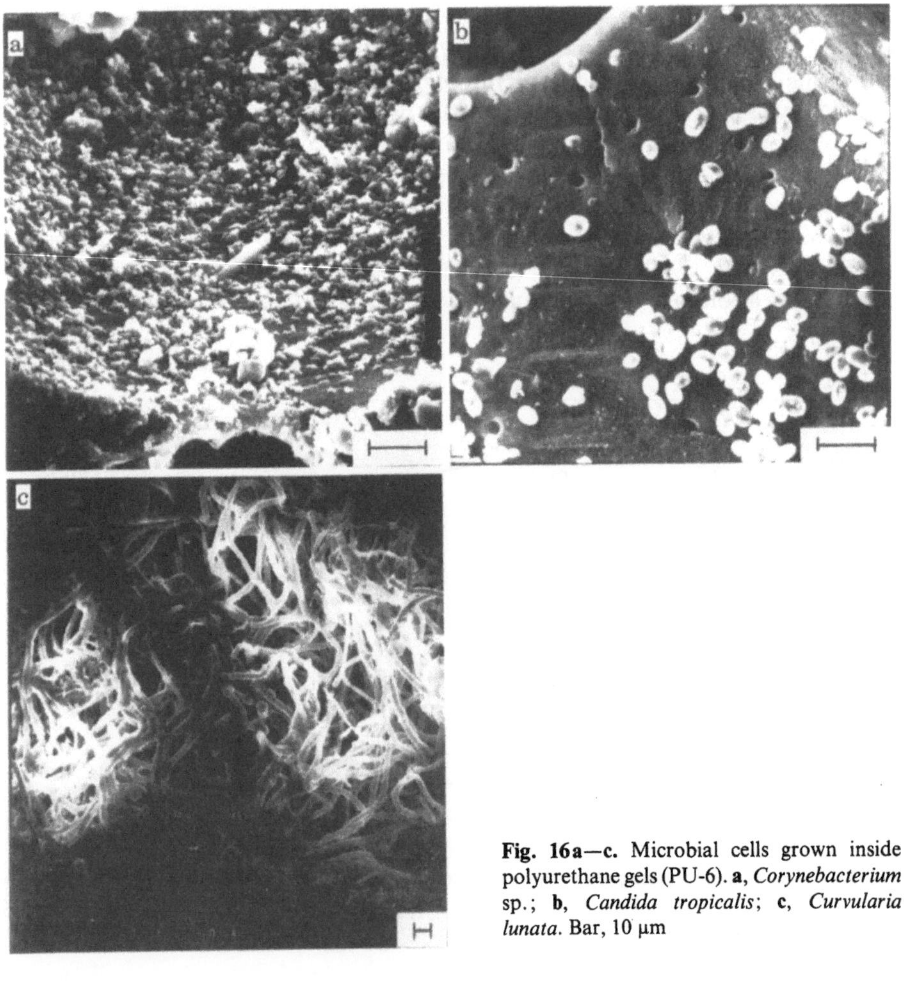

Fig. 16a—c. Microbial cells grown inside polyurethane gels (PU-6). **a**, *Corynebacterium* sp.; **b**, *Candida tropicalis*; **c**, *Curvularia lunata*. Bar, 10 μm

A strain of *Propionibacterium* sp. entrapped with a urethane prepolymer, PU-9, synthesized *de novo* vitamin B_{12}, the most complicated non-polymer organic compound of known structure in nature. When 5 g of wet cells were entrapped with 1 g of PU-9 and cultivated statically in 50 ml of a medium, the entrapped cells accumulated about 1 mg of the vitamin in the medium after 6 batches (each batch, 3 days) [44]. Entrapment of the bacterial cells with a photo-crosslinkable resin prepolymer, ENT-4000, also gave a good result. The growing cells of *Corynebacterium* sp. entrapped in photo-crosslinked gels of a hydrophilic or hydrophobic nature exhibited the ability to hydroxylate 4-androstene-3,17-dione at the 9α-position even in the presence of 15% of dimethyl sulfoxide [55], as described previously.

In the case of fungi, it is very difficult to entrap homogeneously living fungal mycelia and the homogenization of mycelia prior to immobilization often causes the loss of enzyme activity. Furthermore, enzymes leaks from mycelium-entrapped gels when the gels are cut into small pieces. However, fungi are important microorganisms that are used in industry to produce and convert various compounds. Hence, the entrapment of fungal spores and subsequent cultivation into the mycelial form were attempted just recently [3, 89].

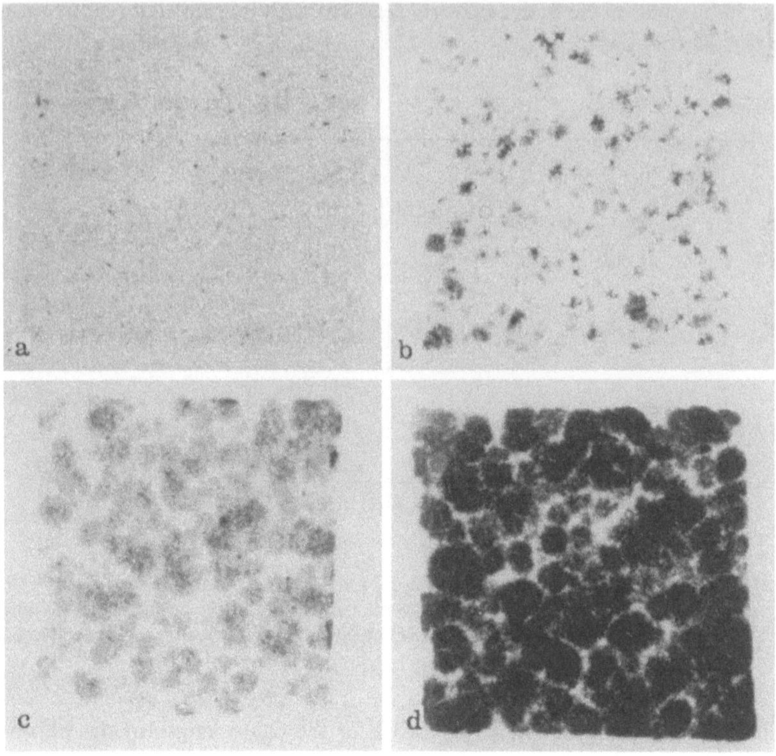

Fig. 17a—d. Growth of *Curvularia lunata* inside gel matrices. [3,52]. Spores of *C. lunata* were entrapped with a photo-crosslinkable resin prepolymer, ENT-4000 (**a**), and were incubated for 28 h (**b**), 48 h (**c**), and 96 h (**d**) in a nutrient medium. Size of each gel, 5 × 5 mm

Spores of *Curvularia lunata* were entrapped by different methods. When the entrapped spores were incubated in a nutrient medium containing cortexolone as inducer, the activity of the steroid 11β-hydroxylation system, which converts cortexolone to hydrocortisone (see, Fig. 4), was induced with the germination of spores and development of mycelia [50, 52]. Among various entrapment methods, entrapment with photo-crosslinkable resin prepolymers of appropriate chain-length was found to be the best for the development of mycelia, and, subsequently, for the activity of the resulting gel-entrapped mycelia to hydroxylate cortexolone. On the other hand, germination of the spores was rather inhibited when they were entrapped with the urethane prepolymers, PU-3 and PU-6. The size of the gel matrices — that is, chain-length of the prepolymers — had an important effect on the development of mycelia in gels (see, Fig. 3). As regards the development of mycelia and the induction of the hydroxylation activity, ENT-4000 was the most suitable among the hydrophilic photo-crosslinkable resin prepolymers examined. The growth of the cells inside ENT-4000 gels can be clearly observed in Fig. 17. The entrapped mycelia were found to be far more stable than the free counterparts and could be reactivated by incubation of the gels in a nutrient medium containing cortexolone as inducer. A typical example is shown in Fig. 18 when the entrapped mycelia were used repeatedly for 50 batches of the hydroxylation reaction with intermittent reactivation (operational period, 100 days).

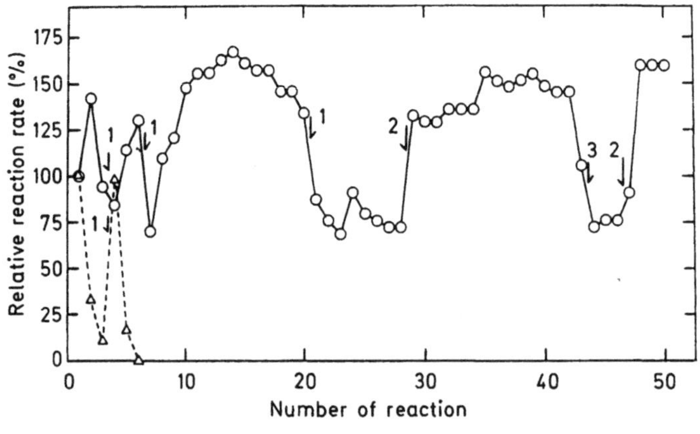

Fig. 18. Repeated use of *Curvularia lunata* mycelia in the hydroxylation of cortexolone [51,52]. Each reaction was carried out for 48 h. Arrows indicate the incubation of free or entrapped mycelia in a nutrient medium under various conditions. (○), ENT-4000-entrapped mycelia; (△), free mycelia

A similar effect of the size of the gel net-work or the chain-length of the photo-crosslinkable resin prepolymers was observed on the development of *Rhizopus stolonifer* mycelia and subsequent induction of the steroid 11α-hydroxylation system for converting progesterone to 11α-hydroxyprogesterone [54]. In this case, an equal weight mixture of ENT-2000 and ENT-4000 was found to be the most suitable of all the prepolymers tested.

Pilot-scale production of ethanol has been successfully carried out by using growing *Saccharomyces* cells entrapped with photo-crosslinkable resin prepolymers [91]. The system has been operated for long periods with a high yield of ethanol.

As mentioned above, not only bacterial cells and yeast cells but also fungal spores can be entrapped under living conditions with photo-crosslinkable resin prepolymers or by urethane prepolymers. In these cases, however, selection of suitable prepolymers, depending on the kind of microorganism, is very important for obtaining entrapped preparations with excellent properties.

7 Future Prospects

At present, the utilization of the prepolymer methods mentioned here is very limited because such prepolymers are not commercially available. However, these methods, which enable gel properties to be easily selected, seem to be indispensable for achieving useful bioreactions of the second generation. For example, the introduction of biochemical processes into chemical industries is attracting worldwide attention. For this purpose, synthesis and conversion of a variety of compounds must be carried out in the presence of organic solvents, or it should be said that substrates themselves are organic solvents toward which biocatalysts have to maintain their activities. In these cases, the properties of the gels which entrap biocatalysts will seriously affect the efficiency of the reactions. The prepolymers utilized hitherto were too few in number to be able to cover all the types of bioconversions. New types of prepolymers, which are easy to handle and whose gel properties can be easily controlled, should be developed for the extensive application of prepolymer methods to bioindustries. An ultimate target will be to obtain gels which have quite different affinities for substrates and products.

8 Abbreviations

4-AD	4-androstene-3,17-dione
ADD	1,4-androstadiene-3,17-dione
ADP	adenosine diphosphate
AMP	adenosine monophosphate
ATP	adenosine triphosphate
CDP	cytidine diphosphate
CMP	cytidine monophosphate
DHEA	dehydroepiandrosterone
DTS	Δ^1-dehydrotestosterone
ENT	hydrophilic photo-crosslinkable resin prepolymers, whose main chain is poly(ethylene glycol)
ENTA	anionic photo-crosslinkable resin prepolymers
ENTB	hydrophobic photo-crosslinkable resin prepolymers with polybutadiene as main chain
ENTE	emulsion type photo-crosslinkable resin prepolymers

ENTP	hydrophobic photo-crosslinkable resin prepolymers, whose main chain is poly(propylene glycol)
NAD	nicotinamide adenine dinucleotide
NADP	nicotinamide adenine dinucleotide phosphate
PAAM	polyacrylamide
PAN	poly(acrylamide-co-*N*-acryloxysuccinimide)
PB-200k	photo-crosslinkable resin prepolymers, whose main skeletons are
PBM-2000	polybutadiene and maleic polybutadiene, respectively
PEGA	poly(ethylene glycol) diacrylate
PEGM	poly(ethylene glycol) dimethacrylate
PMS	phenazine methosulfate
PU	urethane prepolymers
PVA	polyvinyl alcohol
TS	testosterone

9 References

1. Methods in Enzymology, Vol. 44, (Mosbach, K., ed.), New York: Academic Press 1976
2. Immobilized Enzyme. Research and Development, (Chibata, I., ed.), Tokyo: Kodansha and New York: John Wiley and Sons 1978
3. Fukui, S., Tanaka, A.: Ann. Rev. Microbiol. *36*, 145 (1982)
4. Fukui, S. et al.: FEBS Lett. *66*, 179 (1976)
5. Tanaka, A. et al.: J. Ferment. Technol. *55*, 71 (1977)
6. Chun, Y. Y., Iida, M., Iizuka, H.: J. Gen. Appl. Microbiol. *27*, 505 (1981)
7. Tanaka, A. et al.: Eur. J. Biochem. *80*, 193 (1977)
8. Tanaka, A. et al.: J. Ferment. Technol. *56*, 511 (1978)
9. Sonomoto, K. et al.: Eur. J. Appl. Microbiol. Biotechnol. *6*, 325 (1979)
10. Omata, T. et al.: ibid. *8*, 143 (1979)
11. Omata, T. et al.: ibid. *6*, 207 (1979)
12. Fukui, S., Tanaka, A., Gellf, G.: Enzyme Engineering *4*, 299 (1978)
13. Fukui, S. et al.: Biochimie *62*, 381 (1980)
14. Itoh, N. et al.: J. Appl. Biochem. *1*, 291 (1979)
15. Miyairi, S.: Biochim. Biophys. Acta *571*, 374 (1979)
16. Fukushima, S. et al.: Biotech. Bioeng. *20*, 1465 (1978)
17. Sonomoto, K. et al.: Agric. Biol. Chem. *44*, 1119 (1980)
18. Tanaka, A. et al.: Eur. J. Appl. Microbiol. Biotechnol. *7*, 351 (1979)
19. Klein, J., Kluge, M.: Biotechnol. Lett. *3*, 65 (1981)
20. Fusee, M. C., Swann, W. E., Calton, G. J.: Appl. Environ. Microbiol. *42*, 672 (1981)
21. Maeda, H., Suzuki, H., Yamauchi, A.: Biotech. Bioeng. *15*, 607 (1973)
22. Tanaka, Y. et al.: ibid. *24*, 857 (1982)
23. Yoshii, F., Fujimura, T., Kaetsu, I.: ibid. *23*, 833 (1981)
24. Freeman, A., Aharonowitz, Y.: Abstr. VIth Int. Ferment. Symp., p. 121 (1980)
25. Freeman, A., Aharonowitz, Y.: Biotech. Bioeng. *23*, 2747 (1981)
26. Pollak, A. et al.: J. Amer. Chem. Soc. *100*, 302 (1978)
27. Sakata, K., Kitano, H., Ise, N.: J. Appl. Biochem. *3*, 518 (1981)
28. Klein, J., Wagner, F.: Dechema Monographs *82*, 142 (1978)
29. Vorlop, K. D., Klein, J., Wagner, F.: Abstr. VIth Int. Ferment. Symp., p. 122 (1980)
30. Baughn, R. L., Adalsteinsson, Ö., Whitesides, G. M.: J. Amer. Chem. Soc. *100*, 304 (1978)
31. Yokozeki, K. et al.: Eur. J. Appl. Microbiol. Biotechnol. *14*, 1 (1982)
32. Yokozeki, K. et al.: Enzyme Engineering *6*, 151 (1982)
33. Kimura, Y. et al.: Eur. J. Appl. Microbiol. Biotechnol. *17*, 107 (1983)
34. Egerer, P. et al.: Biotechnol. Lett. *4*, 489 (1982)
35. Tanaka, A. et al.: J. Ferment. Technol. *58*, 391 (1980)

36. Watanabe, K. et al.: Agric. Biol. Chem. *46*, 119 (1982)
37. Watanabe, K., Tanaka, A., Fukui, S.: ibid. *46*, 289 (1982)
38. Asada, M. et al.: ibid. *43*, 1773 (1979)
39. Kimura, A. et al.: Eur. J. Appl. Microbiol. Biotechnol. *5*, 13 (1978)
40. Kimura, A. et al.: ibid. *11*, 78 (1981)
41. Yokozeki, K. et al.: ibid. *14*, 225 (1982)
42. Klein, J., Wagner, F.: Enzyme Engineering *5*, 335 (1980)
43. Tanaka, A., Itoh, N., Fukui, S.: Agric. Biol. Chem. *46*, 127 (1982)
44. Yongsmith, B. et al.: Eur. J. Appl. Microbiol. Biotechnol. *16*, 70 (1982)
45. Omata, T., Tanaka, A., Fukui, S.: J. Ferment. Technol. *58*, 339 (1980)
46. Fukui, S. et al.: Enzyme Engineering *5*, 347 (1980)
47. Fukui, S. et al.: Eur. J. Appl. Microbiol. Biotechnol. *10*, 289 (1980)
48. Fukui, S., Tanaka, A.: Acta Biotechnol. *1*, 339 (1981)
49. Yamane, T. et al.: Biotech. Bioeng. *21*, 2133 (1979)
50. Sonomoto, K. et al.: J. Ferment. Technol. *59*, 465 (1981)
51. Tanaka, A. et al.: Enzyme Engineering *6*, 131 (1982)
52. Sonomoto, K. et al.: Appl. Environ. Microbiol. *45*, 436 (1983)
53. Fukui, S. et al.: unpublished
54. Sonomoto, K. et al.: Eur. J. Appl. Microbiol. Biotechnol. *16*, 57 (1982)
55. Sonomoto, K. et al.: ibid. *17*, 203 (1983)
56. Omata, T. et al.: ibid. *11*, 199 (1981)
57. Tanaka, A., Fukui, S.: Immobilized Cells and Organelles, Vol. 1, p. 101, (Mattiasson, B., ed.), Boca Raton: CRC Press 1983
58. Ochiai, H., Tanaka, A., Fukui, S.: Appl. Biochem. Bioeng. *4*, 153 (1983)
59. Larreta Garde, V. et al.: Eur. J. Appl. Microbiol. Biotechnol. *11*, 133 (1981)
60. Cocquempot, M. F. et al.: ibid. *11*, 193 (1981)
61. Tanaka, A. et al.: Agric. Biol. Chem. *44*, 2399 (1980)
62. Fukui, S., Tanaka, A.: J. Appl. Biochem. *1*, 171 (1979)
63. Tanaka, A. et al.: Eur. J. Appl. Microbiol. Biotechnol. *5*, 17 (1978)
64. Fukui, S., Tanaka, A.: Enzyme Engineering *6*, 191 (1982)
65. Martinek, K., Semenov, A. N., Berezin, I. V.: Biochim. Biophys. Acta *658*, 76 (1981)
66. Martinek, K., Semenov, A. N.: ibid. *658*, 90 (1981)
67. Martinek, K., Berezin, I. V.: J. Solid-Phase Biochem. *2*, 343 (1977)
68. Klibanov, A. M.: Anal. Biochem. *93*, 1 (1979)
69. Butler, L. G.: Enzyme Microb. Technol. *1*, 253 (1979)
70. Antonini, E., Carrea, G., Cremonesi, P.: ibid. *3*, 291 (1981)
71. Ohlson, S., Larsson, P. O., Mosbach, K.: Biotech. Bioeng. *20*, 1267 (1978)
72. Yang, H. S., Studebaker, J. F.: ibid. *20*, 17 (1978)
73. Maddox, I. S., Dunnill, P., Lilly, M. D.: ibid. *23*, 345 (1981)
74. Atrat, P., Groh, H.: Z. Allg. Mikrobiol. *21*, 3 (1981)
75. Ingalls, R. G., Squires, R. G., Butler, L. G.: Biotech. Bioeng. *17*, 1627 (1975)
76. Homandberg, G. A., Mattis, J. A., Laskowski, M., Jr.: Biochemistry *17*, 5220 (1978)
77. Yamashita, M. et al.: J. Agr. Food Chem. *23*, 27 (1975)
78. Yamashita, M., Arai, S., Fujimaki, M.: J. Food Sci. *41*, 1029 (1976)
79. Morihara, K., Oka, T., Tsuzuki, H.: Nature *280*, 412 (1979)
80. Morihara, K. et al.: Biochem. Biophys. Res. Commun. *92*, 396 (1980)
81. Jonczyk, A., Gattner, H.-G.: Hoppe-Seyler's Z. Physiol. Chem. *362*, 1591 (1981)
82. Becker, W., Pfeil, E.: J. Amer. Chem. Soc. *88*, 4299 (1966)
83. Butler, L. G., Reithel, F. J.: Arch. Biochem. Biophys. *178*, 43 (1977)
84. Utagawa, T. et al.: FEBS Lett. *109*, 261 (1980)
85. Klibanov, A. M. et al.: Biotech. Bioeng. *19*, 1351 (1977)
86. Atrat, P., Hüller, E., Hörhold, C.: Z. Allg. Mikrobiol. *20*, 79 (1980)
87. Tanaka, T. et al.: Agric. Biol. Chem. *45*, 2387 (1981)
88. Chibata, I., Tosa, T.: Trends Biochem. Sci. *5*, 88 (1980)
89. Ohlson, S. et al.: Eur. J. Appl. Microbiol. Biotechnol. *10*, 1 (1980)
90. Asada, M. et al.: Agric. Biol. Chem. *46*, 1687 (1982)
91. Oda, G., Samejima, H., Yamada, T.: Proc. Biotech *83*, 597 (1983)

Regulation of Respiration and Its Related Metabolism by Vitamin B_1 and Vitamin B_6 in *Saccharomyces* Yeasts

Teijiro Kamihara and Ichiro Nakamura
Laboratory of Industrial Biochemistry, Department of Industrial Chemistry, Faculty of Engineering, Kyoto University, Kyoto 606, Japan

Thiamine added to a vitamin B_6-free medium was accumulated in the free form in *Saccharomyces* yeast cells, and caused growth inhibition and respiratory deficiency under aerobic conditions. The thiamine-induced respiratory deficiency occurred as the result of a sequence of events as follows: (1) a decrease in vitamin B_6 content, (2) a reduction in the activity of δ-aminolevulinate synthase, (3) heme deficiency, (4) cytochrome deficiency, and (5) respiratory deficiency. However, the growth inhibition was shown to be partially due to the respiratory deficiency, and the participation of some other events caused by the thiamine-induced vitamin B_6 deficiency was suggested. The cytochrome deficiency also caused alteration in lipid composition; unsaturated fatty acid content was decreased and sterol composition was changed. Associated with the respiratory deficiency, the activity of NAD-linked glutamate dehydrogenase was decreased. In contrast, the NADP-linked enzyme activity was markedly increased. It was suggested that this increase was not caused by the respiratory deficiency but by a thiamine-enhanced glucose effect. The thiamine-grown cells showed an altered amino acid pool. Reflecting the respiratory deficiency, the thiamine-grown cells showed elevated activities of glycolysis and ethanol production. It was, however, suggested that the glycolytic rate was not determined by phosphofructokinase activity, and ethanol production was controlled independently of glycolytic activity in the thiamine-supplemented culture.

1 Introduction

Aerobically grown cells of *Saccharomyces* yeasts possess matured and developed mitochondria with significantly high activity of respiration. In contrast, yeast cells grown under anaerobic conditions have atrophied or undifferentiated mitochondria named promitochondria which have no cristae and hence no respiratory activity. Upon aeration of the anaerobically grown culture, respiratory competence is immediately attained with development of mitochondria. A similar effect of aeration can be observed without cell proliferation [12]. This phenomenon is known as "respiratory adaptation" and widely utilized for the study of mitochondrial biogenesis as a good experimental system.

For energy formation, *Saccharomyces* yeasts depend rather upon ethanol production via glycolysis even under aerobic conditions. However, glycolytic and ethanol-producing activities of the yeasts are also influenced by respiratory competence. When cells are growing under aerobic conditions, both glucose utilization and ethanol production are relatively lowered. The mechanism of this well-known phenomenon, the Pasteur Effect, has been investigated extensively in *Saccharomyces* yeasts as well as in bacterial and mammalian cells [11, 13, 41, 72].

Even under aerobic conditions, the presence of high concentrations of glucose suppresses respiration (Crabtree Effect) [10]. Yeast growth depends exclusively on ethanol production under these conditions. Similar effects are observed with some respiratory inhibitors such as chloramphenicol [46] and α-phenylethyl alcohol [33].

We have found that thiamine ($1\ \mu g\ ml^{-1}$) also causes respiratory depression in *Saccharomyces* yeasts even under aerobic conditions when added to vitamin B_6-free medium, and pyridoxine ($0.02\ \mu g\ ml^{-1}$) prevents the effect of thiamine [51, 52]. This would be for the first time that such physiologically active substances have been shown to be effective for respiratory activity of yeast cells at the physiologically significant concentrations.

On the other hand, it is well known that *Saccharomyces carlsbergensis* 4228 requires vitamin B_6 compounds for growth when thiamine is added to a synthetic medium [3]. This phenomenon has been utilized for a microbiological assay of vitamin B_6. Rabinowitz et al. [71] demonstrated that vitamin B_6 compounds are required for preventing the thiamine-induced growth inhibition. There are many reports [6, 35, 36, 64] dealing with the inhibitory effect of thiamine and the preventing action of vitamin B_6 compounds on the growth of this yeast. However, unequivocal explanation for the mechanism had not been presented when we discovered the thiamine-induced respiratory deficiency. Our finding provided a partial but significant solution for this long-pending problem as described below.

This paper summarizes (1) the mechanism of the thiamine-induced respiratory deficiency, (2) the relation of the respiratory deficiency to the growth depression, and (3) alteration in some important metabolism which are related to respiration.

2 Effects of Thiamine and Pyridoxine on Respiratory Activity of *Saccharomyces carlsbergensis*

2.1 Respiratory Deficiency Caused by Thiamine and Its Prevention by Pyridoxine [51, 53, 55]

Saccharomyces carlsbergensis 4228 [ATCC 9080] was grown aerobically at 30 °C in a defined medium containing 5% glucose and 0.4% casamino acids. Thiamine hydrochloride (1 μg ml^{-1}) and pyridoxine hydrochloride (0.02 μg ml^{-1}) were added to the medium when indicated. The cells grown with and without added thiamine are designated as "thiamine-grown cells" and "control cells", respectively. The cells grown with both the vitamins are called "thiamine and pyridoxine-grown cells".

The addition of thiamine to the growth medium caused a delay of the beginning of growth, a decrease in growth rate and a lowered maximum level of growth under aerobic conditions (Fig. 1). The lowering of maximum level of growth suggested a decrease in energy efficiency in the thiamine-grown cells. The growth inhibition by thiamine was prevented completely by the addition of pyridoxine.

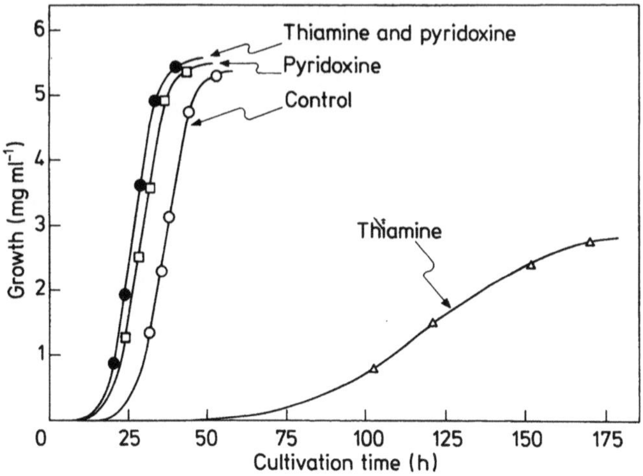

Fig. 1. Effects of thiamine and pyridoxine on the growth of *Saccharomyces carlsbergensis* 4228[51]. Cultivation was carried out as described in the text

Figure 2 shows the time-course of respiration rate of the cells, the growth curves of which are given in Fig. 1. The oxygen uptake by the thiamine-grown cells was markedly low at any phase of growth. Pyridoxine added to the growth medium, singly or concomitantly with thiamine, gave normal respiratory competence to the yeast. The low respiration rate in the thiamine-grown cells is consistent with the low maximum level of growth. The respiratory deficiency may not be due to glucose effect, because residual glucose concentration of the thiamine-grown culture was not higher at any growth phase than that of the control or the thiamine and pyridoxine-grown culture. There have been no reports dealing with respiratory depression caused by

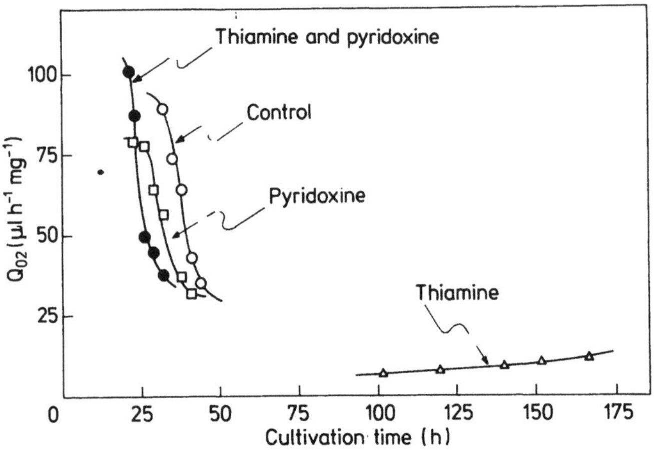

Fig. 2. Effects of thiamine and pyridoxine on the respiration rate of intact cells [51]. Cells harvested at the indicated time of incubation were washed and suspended in 0.05 M citrate buffer (pH 5.3). An aliquot of the suspension containing 6 mg of cells was transferred to a Warburg flask. Glucose (17 μmol) was added as substrate from a side arm. Oxygen uptake was measured every 5 min for 40–60 min. The respiratory rate of the cells was calculated as Q_{O_2} (μl O_2 h^{-1} mg^{-1})

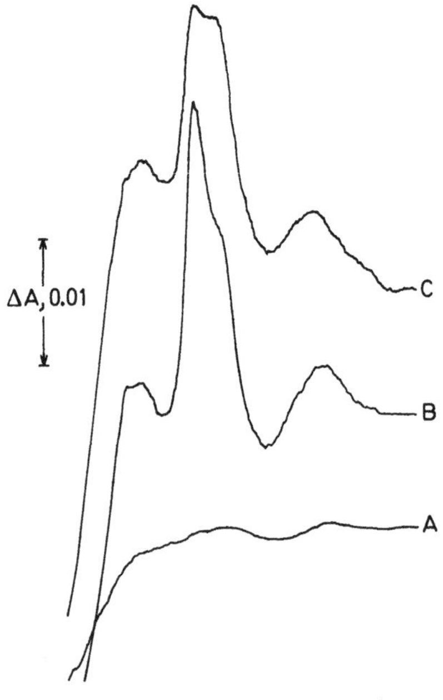

Fig. 3. Cytochrome spectra of cells [53]. Cells were harvested when the cell yield reached to 0.5 to 0.7 mg of dry cells per ml. The spectra were taken at room temperature with cell concentrations of 12.7, 12.9 and 13.8 mg of dry cells per ml for the thiamine-grown cells (A), the control cells (B), and the thiamine and pyridoxine-grown cells (C), respectively. ΔA: change in absorbance

physiologically active substances other than glucose. The results obtained clearly indicated that thiamine added exogenously to the growth medium caused a remarkable reduction of the respiratory activity under aerobic conditions. The prevention of this effect of thiamine by concomitant addition of pyridoxine suggested a close relationship of this phenomenon to the well-known inhibitory effect of thiamine on the growth of the yeast and its prevention by vitamin B_6 compounds.

The absorption spectra of the yeast cells are presented in Fig. 3. The control cells showed the characteristic cytochrome peaks corresponding to cytochrome aa_3, b and c at 605, 563 and 552 nm, respectively, whereas the thiamine-grown cells had no distinguishable peaks in the region. This effect of thiamine was also prevented by the concomitant addition of pyridoxine to the medium. These observations are consistent with the decrease in the rate of respiration in the thiamine-grown cells and its prevention by pyridoxine.

The above-mentioned effects of thiamine and pyridoxine on respiratory activity of the yeast cells were further confirmed by experiments using cell-free extracts. As shown in Fig. 4, the extracts of the thiamine-grown cells exhibited no appreciable activities of respiratory chain enzymes throughout the growth period. Elongation of cultivation time caused no significant increase in these enzyme activities. On the contrary, the levels of the enzymes in the thiamine and pyridoxine-grown cells were

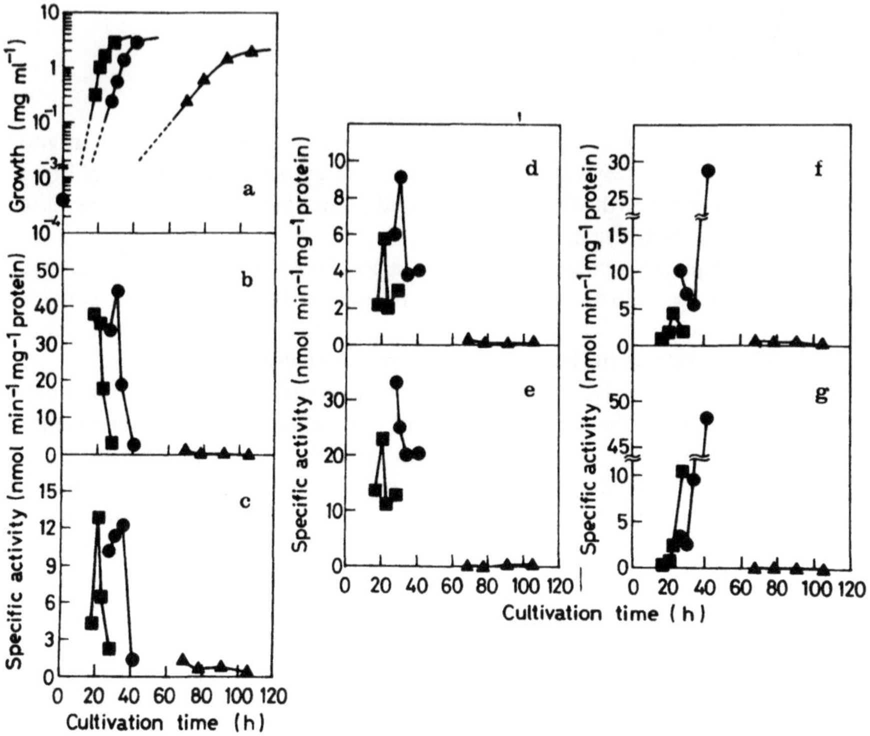

Fig. 4a—g. Cellular activities of respiratory chain enzymes and other heme-containing enzymes [53]. Cell-free extracts were prepared, and the enzymes were assayed spectrophotometrically as described in Ref. 53. (●) Control cells, (▲) thiamine-grown cells, (■) thiamine and pyridoxine-grown cells. **a** Growth; **b** cytochrome oxidase; **c** NADH oxidase, **d** succinate-cytochrome c oxidoreductase, **e** NADH-cytochrome c oxidoreductase; **f** Lactate dehydrogenase; **g** catalase. In section **a** the closed circle at zero time of cultivation indicates the inoculum size

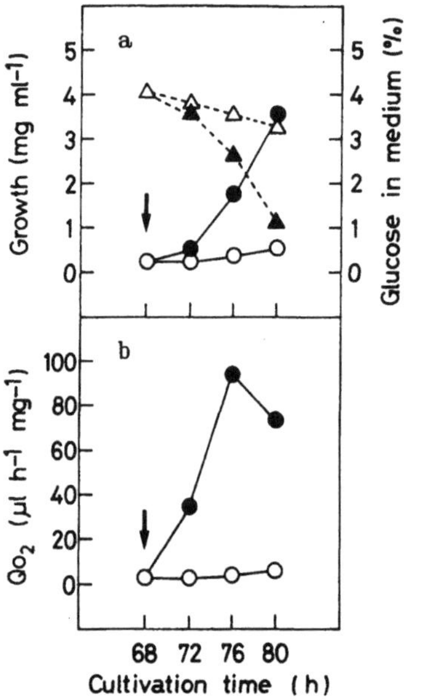

Fig. 5a and b. Comparison of growth, glucose concentration and respiratory activity in the presence and absence of pyridoxine during cultivation with thiamine [53]. Cells were grown with thiamine for 68 h to the late exponential phase. Pyridoxine was then added at a final concentration of 0.02 µg per ml. The addition of pyridoxine is indicated by an arrow. **a** Growth of cells: (○) without pyridoxine, (●) with pyridoxine. Glucose concentration: (△) without pyridoxine, (▲) with pyridoxine; **b** Respiratory activity (Q_{O_2}): (○) without pyridoxine, (●) with pyridoxine

high and almost the same as those in the control cells. These results were in good agreement with those of respiration in Fig. 2 and of cytochrome spectra in Fig. 3, and suggested that the growth stimulation by pyridoxine would be due to the increased activities of respiratory enzymes.

Not only the respiratory chain enzymes but also other heme-containing enzymes, lactate dehydrogenase and catalase, were also influenced by thiamine. The activities of these enzymes were also negligible in the thiamine-grown cells, but increased to some significant extents in the thiamine and pyridoxine-grown cells.

Pyridoxine was effective also when added in the course of cultivation with thiamine. Pyridoxine was added to the culture which had been grown to the late exponential phase. Figure 5 shows the changes in growth and glucose concentration and in respiratory activity of cells after the addition of pyridoxine. Both growth and glucose consumption were enhanced but the extent was small at the initial stage of incubation (4 h). In contrast, respiratory activity increased even in the period to a level more than one third of a maximum value which was attained after 8 h incubation. It was, therefore, suggested that the occurrence of normal growth was a result of the pyridoxine-induced restoration of respiratory activity. Associated with the increase in respiratory activity, the activities of respiratory chain enzymes and other heme-containing enzymes were also increased (Fig. 6). The effect of pyridoxine occurred almost immediately after its addition as in the case of respiratory activity. Also, normal cytochrome spectra appeared within 4 h upon the addition of pyridoxine.

The above finding that all the heme-containing enzymes tested were influenced by thiamine suggested the presence of a mechanism by which heme biosynthesis is inhibited or its degradation is stimulated in the presence of added thiamine. A clue to this phenomenon was offered by the fact that the vitamin B_6 content of the thiamine-grown cells was substantially low throughout the cultivation period as

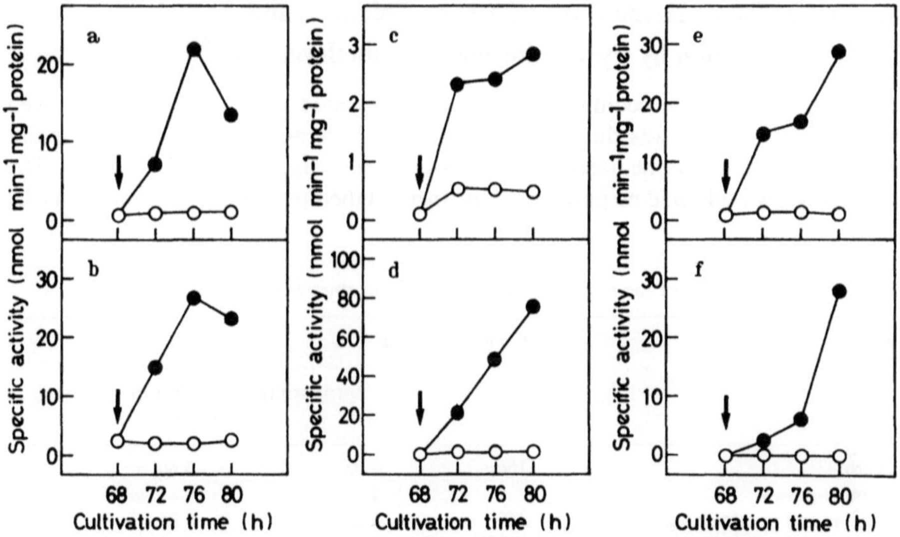

Fig. 6a—f. Comparison of the activities of heme-containing enzymes in the presence and absence of pyridoxine during cultivation with thiamine [53] The experimental details are described in the legend to Fig. 5 The addition of pyridoxine is indicated by an arrow (○) Without pyridoxine; (●) with pyridoxine. **a** Cytochrome oxidase; **b** NADH oxidase; **c** succinate-cytochrome c oxidoreductase; **d** NADH-cytochrome c oxidoreductase; **e** lactate dehydrogenase; **f** catalase

compared with that of the control cells (Table 1). The vitamin B_6 deficiency was regarded as the primary effect of thiamine and the cause of heme deficiency. This consideration was supported by the effect of pyridoxine protecting the yeast cells from the thiamine-induced respiratory deficiency. The requirement of pyridoxal phosphate for heme biosynthesis was first observed by nutritional experiments [78, 90], and now established at the step of δ-aminolevulinate (ALA) synthesis [21, 22, 39, 43, 44]. The thiamine-induced deficiency in vitamin B_6 was considered to cause a decrease in the activity of ALA synthase catalyzing this step, and consequently the reduction of heme synthesis. Accumulation of some porphyrin-related substances reported in a respiratory deficient mutant of *S. cerevisiae* [81], or that of abnormal hemoproteins was not observed in the thiamine-grown cells as judged by the absorption spectra of cell suspension.

Table 1. Effect of thiamine on the cellular content of vitamin B_6

Growth phase	Control cells		Thiamine-grown cells	
	Vitamin B_6 content[a]	Cell growth[b]	Vitamin B_6 content[a]	Cell growth[b]
Late exponential	11.5	0.20	0.61	0.16
Early stationary	6.6	1.53	1.58	0.55
Stationary	2.6	2.75	0.77	1.76

[a] ng pyridoxine · HCl per mg dry cells, determined by microbioassay using *S. carlsbergensis* 4228 [53],
[b] dry cells, mg ml^{-1}

2.2 Depression of Heme Biosynthesis by Thiamine and Its Restoration by Pyridoxine and δ-Aminolevulinate [55]

As described in the preceding section, it was suggested that the thiamine-induced respiratory deficiency was due to decreased heme biosynthesis. To prove this point the content of the intermediates in heme biosynthesis was determined (Table 2). The thiamine-grown cells had no porphyrins, neither did their precursor, ALA, whereas the

Table 2. Cellular content of the intermediates in the heme-synthesizing pathway

Cells	Content (nmol g^{-1}) a		
	ALA	Coproporphyrin III	Protoporphyrin IX
Control	150	0.02	0.60
Thiamine-grown	0	0	0
Thiamine and pyridoxine-grown	240	0.10	0.85
Thiamine and ALA-grown	ND[b]	0.08	0.64

[a] Cells were harvested at the late exponential growth phase. ALA and the porphyrins were assayed as described in Ref. [55]. Porphobilinogen was not detectable, not only in the thiamine-grown cells but also in the control cells and the thiamine and ALA-grown cells;

[b] Not determined;

Experimental details are described in Ref. [55]

control cells contained significant amounts of these substances. The results suggested that the supply of ALA was blocked in the thiamine-grown cells. This was evidenced by the effect of ALA added concomitantly with thiamine to the culture medium (the cells grown with thiamine and ALA are termed "thiamine and ALA-grown cells"). The addition of ALA caused the occurrence of porphyrins in the cells. ALA and the porphyrins were present at high concentrations also in the thiamine and pyridoxine-grown cells which possess normal respiratory activity.

The effect of pyridoxine suggested that the absence of ALA in the thiamine-grown cells was due to vitamin B_6 deficiency induced by thiamine. Porphobilinogen was not detected not only in the thiamine-grown cells but also in the pyridoxine or ALA supplemented cells.

The effect of ALA on porphyrin biosynthesis was further demonstrated by measuring the cellular activities of heme-containing enzymes, and also by taking cytochrome spectra in cell suspensions. As shown in Fig. 7, the heme-containing enzyme activities were extremely low in the thiamine-grown cells throughout the cultivation period. The thiamine and ALA-grown cells, as well as the control cells, exhibited variations in the enzyme activities in the course of cultivation. However, the values were higher than those for the thiamine-grown cells and comparable to the normal levels observed in the control cells. The characteristic absorption peaks of cytochrome aa_3, b, and c were observed in the thiamine and ALA-grown cells, as well as in the thiamine and pyridoxine-grown cells and the control cells. Similar to pyridoxine, ALA also caused immediate restoration of the levels of the heme-containing

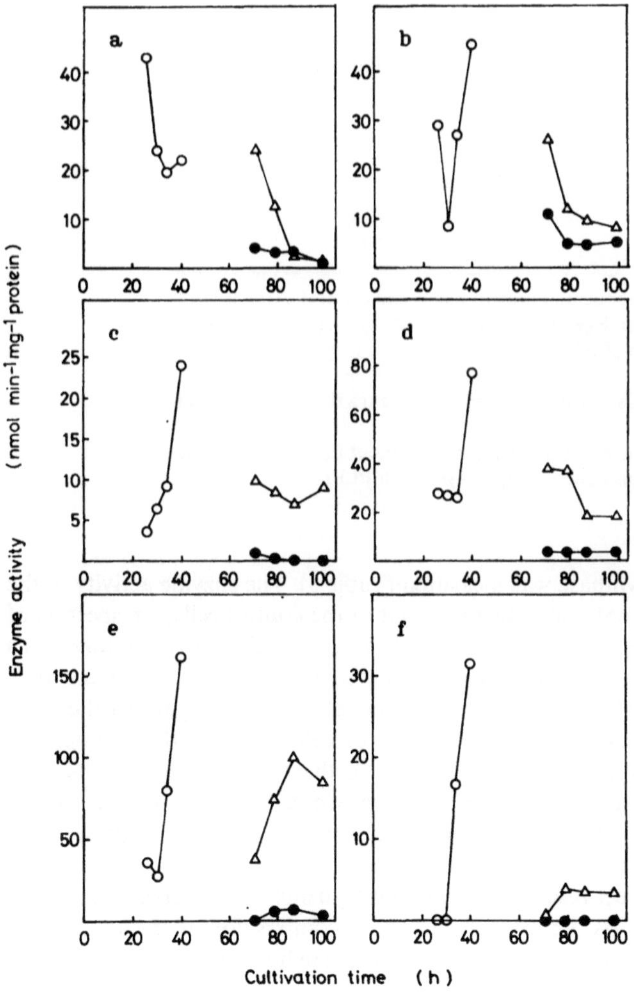

Fig. 7a—f. Effects of ALA on the cellular activities of respiratory chain enzymes and other heme-containing enzymes [55]. Cells were harvested at the indicated times. Enzymes were assayed in cell-free extracts as described in the legend to Fig. 4. (○) Control cells, (●) thiamine-grown cells, (△) thiamine and ALA-grown cells. **a** Cytochrome oxidase; **b** NADH oxidase; **c** succinate-cytochrome c oxidoreductase; **d** NADH-cytochrome c oxidoreductase; **e** lactate dehydrogenase; **f** catalase

enzymes and cytochromes when added to the thiamine-supplemented culture at the exponential growth phase.

The decreased content of vitamin B_6 was not affected by the addition of ALA, indicating that vitamin B_6 deficiency precedes ALA deficiency in the thiamine-grown cells. The absence of ALA in the thiamine-grown cells and the dramatic effect of its addition on the cellular content of porphyrins and of heme-containing enzymes strongly suggested that the thiamine-grown cells might lack a metabolic step(s) essential for the supply of ALA to the heme-biosynthesizing pathway. Hence, the

Table 3. Cellular activity of ALA synthase[a]

Cells	Specific activity[b] of cells assayed		Ratio of holoenzyme to total enzyme[d]
	Without PLP	With PLP[c]	
Control	1.5	2.2	0.68
Thiamine-grown	0.35	0.55	0.64

[a] Cells were harvested at the exponential growth phase and cell-free extracts were prepared as described in the legend to Fig. 4. ALA synthase was assayed by determining the formation of ALA [55];
[b] nmol h^{-1} mg^{-1} protein;
[c] Pyridoxal phosphate (PLP) was added to the assay mixture at a final concentration of 150 µM;
[d] Ratio of specific activity of the holoenzyme (assayed without PLP) to that of the total enzyme (holoenzyme plus apoenzyme, assayed with PLP)

cellular activity of ALA synthase was measured (Table 3). The enzyme activity in the thiamine-grown cells was only one-fourth of that in the control cells, irrespective of the presence or absence of pyridoxal phosphate added in the assay mixture. This indicated that not only the holoenzyme but the total enzyme (holoenzyme plus apoenzyme) were reduced in these cells. The ratio of the holoenzyme to the total enzyme was almost the same between the control cells and the thiamine-grown cells, reflecting the decrease in the total enzyme of the thiamine-grown cells. The severe reduction of ALA supply in the thiamine-grown cells may have been caused by the decrease of both the holoenzyme and the apoenzyme of ALA synthase.

The lowering of the holoenzyme of ALA synthase in the thiamine-grown cells can be considered to be the direct result of the vitamin B_6 deficiency, whereas that of the apoenzyme may be due to its rapid turnover. The decomposition of the apoenzyme by a specific protease in mitochondria has been reported in human liver from pyridoxine-responsive anemia patients [1, 2].

As described above, the activity of ALA synthase was actually low in the thiamine-grown cells. However, the activity was not negligible, but somewhat significant. This would be consistent with the observation that in some experiments the cells also had low but appreciable activities of heme-containing enzymes (Fig. 7). In contrast, no detectable amounts of ALA and other heme precursors were present in the cells (Table 2). This fact can be explained by assuming the rapid conversion of these precursors to heme compounds. However, it would be also possible to consider that in vivo activity of ALA synthase was much lower than the activity assayed in vitro under optimum conditions. There is a possibility that the activity of ALA synthase was very low owing to the lack of the substrates in the thiamine-grown cells. To clarify whether the substrates are supplied sufficiently in the cells, the glycine content in the soluble fraction of the cells and cellular activity of succinyl CoA synthetase were measured. The glycine content in the thiamine-grown cells (12.0 µmol g^{-1}) was comparable to that in the control cells (15.8 µmol g^{-1}). Succinyl CoA synthetase activity was decreased by added thiamine (Table 4).

Table 4. Cellular activity of succinyl CoA synthetase[a]

Cells	Specific activity (nmol h^{-1} mg^{-1} protein)
Control	200
Thiamine-grown	90

[a] Cells were harvested at the exponential growth phase and assayed as described in Ref. [55]

However, the lowered activity was still much higher than the activity of ALA synthase in the thiamine-grown cells (Table 3, 0.35 nmol h^{-1} mg^{-1} protein), indicating that succinyl CoA synthetase did not determine the rate of ALA supply. Further, abnormal lipid composition of the thiamine-grown cells [50,58,59,60] and of their mitochondria [54] (described below) would offer alternative explanations for the idea that the activity of ALA synthase in vivo would be extremely lower than the activity in vitro. First, it may have been that the enzyme could hardly operate in the thiamine-grown cells owing to the defect of its translocation into the mitochondria having the abnormal lipid composition. In rat liver, ALA synthase is known to be synthesized on cytoplasmic ribosomes and translocated into mitochondria [27,91]. Similar mechanism would exist also in yeast. Secondly, the alteration in mitochondrial membrane would affect the activities of activator and inhibitor of ALA synthase which have been reported to be present in mitochondria from rat liver [80] and guinea pig liver [31], respectively. The activator is known to associate with the membrane [80]. If the yeast cells contain similar effectors, a decrease in the activity of the activator or an increase in the activity of the inhibitor could occur as a result of the alteration in mitochondrial membrane and consequently would cause the depression of ALA synthase in vivo. At present, the in vivo activity of ALA synthase remains to be elucidated.

2.3 Relationship between Respiratory Deficiency and Growth Depression Caused by Thiamine [55]

The experimental results described above clearly demonstrated that respiratory deficiency occurred in the cells of *S. carlsbergensis* 4228 grown in a vitamin B_6-free medium with added thiamine as a result of a sequence of events: (1) thiamine-induced vitamin B_6 deficiency, (2) a decrease in ALA synthase activity, (3) a lowering of heme biosynthesis, (4) heme deficiency, (5) cytochrome deficiency and (6) respiratory deficiency.

To know the contribution of the thiamine-induced respiratory deficiency to the growth depression, the effect of ALA on the growth of the yeast with added thiamine was investigated.

Figure 8 shows that the elongated lag time for growth initiation caused by thiamine was not improved by the addition of ALA. The growth rate and the maximum level of growth were increased, but only partially. The effect of ALA on growth was in striking contrast to that of pyridoxine, the addition of which, as shown in Fig. 1, gave a completely normal growth curve under the same conditions as

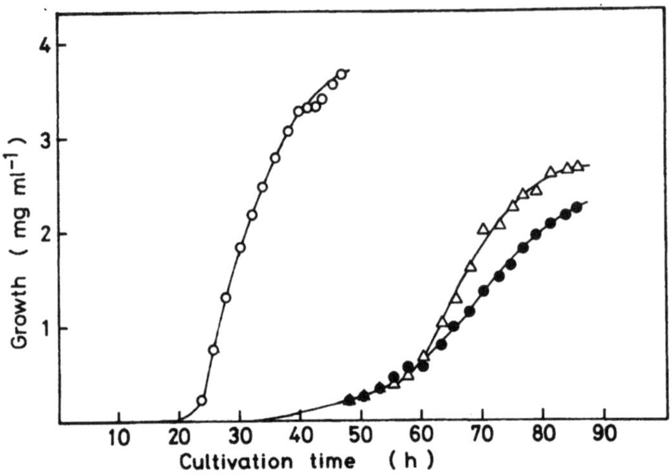

Fig. 8. Effect of ALA on cell growth [55]. Cells were grown with both thiamine · HCl (1 μg ml^{-1}) and ALA · HCl (85 μg ml^{-1}) (△), as well as with thiamine only (●). Cells grown with neither thiamine nor ALA (○) were also measured as a control

those for the addition of ALA. These results suggested that, in addition to the heme deficiency, some other significant events occur in the vitamin B_6-deficient cells. Vitamin B_6 deficiency may bring about alteration in amino acid metabolism and result in a shortage of the cellular content of some amino acids such as tryptophan and cystein which are not contained in the casamino acids used in the culture medium. However, addition of these amino acids to the culture medium had no effect on growth in the presence or absence of ALA. In addition, vitamin B_6 is known to play important roles in the metabolism of carbohydrate, polyamines, and nucleic acids. Metabolic disturbance of these substances may also be responsible for yeast growth under these conditions.

2.4 Respiratory Adaptation Caused by Pyridoxine and by δ-Aminolevulinate in The Thiamine-Grown Cells [52, 55]

2.4.1 Effect of Pyridoxine [52]

The effect of pyridoxine to eliminate the thiamine effects on heme biosynthesis, the activities of heme-containing enzymes, and respiratory activity in *S. carlsbergensis* 4228 cells was observed not only in growing culture but also in resting suspensions of the thiamine-grown cells.

The thiamine-grown cells harvested at the middle exponential growth phase were washed and transferred to a medium (pH 6.3) containing 9.5 mM citric acid, 31 mM potassium citrate, 2% ethanol, 0.1% glucose and 0.02% casamino acids. A heavy cell suspension (4–5 mg ml^{-1}) was used to render cell proliferation during incubation as little as possible. Incubation was carried out aerobically at 30 °C with or without pyridoxine (100 μg ml^{-1}). Glucose and casamino acids were supplied to the medium every 1 h.

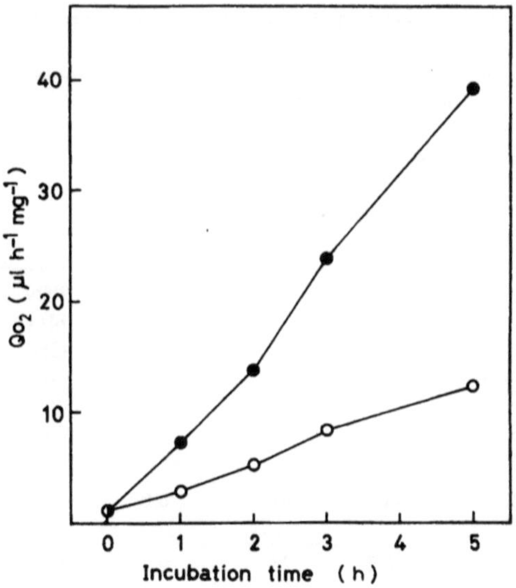

Fig. 9. Increase in the respiration rate by pyridoxine during incubation of the thiamine-grown cells [52]. The incubation of the cells was carried out as described in the text. 5 ml aliquots were withdrawn from the incubation mixture (50 ml) at the indicated time, cooled rapidly, and washed with cold water. The respiration rate of these cells was measured as described in the legend to Fig. 2. (○) Without pyridoxine, (●) with pyridoxine

The respiration rate of the thiamine-grown cells was increased linearly with time during incubation (Fig. 9). Addition of pyridoxine to the medium brought about a 3 to 5-fold increase in the respiratory development. Casamino acids stimulated the pyridoxine-induced respiratory adaptation to a great extent, suggesting the involvement of de novo syntheses of the respiratory enzymes. This possibility was further suggested by experiments with antibiotics. Both chloramphenicol and cycloheximide inhibited completely the increase in the respiration rate (Table 5).

The involvement of pyridoxine in respiratory adaptation was clearly demonstrated in Table 6. The increases in the specific activities of cytochrome-linked respiratory chain enzymes were markedly enhanced by pyridoxine. Thus, the enzyme levels were restored to those in the control cells. The involvement of de novo syntheses of the

Table 5. Effects of chloramphenicol and cycloheximide on the increase in the respiration rate of the thiamine-grown cells during incubation with or without added pyridoxine

Conditions	Additions	Q_{O_2} (µl h^{-1} mg^{-1})
Before incubation	—	6.4
After incubation	None	23.7
	Chloramphenicol	13.9
	Cycloheximide	7.8
	Pyridoxine	109
	Pyridoxine plus chloramphenicol	15.0
	Pyridoxine plus cycloheximide	8.3

Incubation was carried out for 5 h under the same conditions as those in Fig. 9. Chloramphenicol and cycloheximide were added at a final concentration of 3 mg per ml and 3 µg per ml, respectively. For details, see the text and Ref. [52]

Table 6. Increases in the activities of respiratory chain enzymes and other heme-containing enzymes during incubation with pyridoxine

Enzymes	Specific activities (nmol min^{-1} mg^{-1} protein)[a]					
	Control cells	Thiamine-grown cells				
		Before incubation	After incubation with[b]			
			None	PIN	PIN plus CAP	PIN plus CHI
Experiment 1						
Succinate-cytochrome c oxidoreductase	5.5	0.57	2.5	3.5	1.9	0.62
NADH-cytochrome c oxidoreductase	73	3.4	11	61	6.0	5.7
Cytochrome oxidase	35	1.8	6.3	28	0.2	1.2
NADH oxidase	30	5.9	6.4	29	4.3	5.8
Experiment 2						
Lactate dehydrogenase	49	9.8	15	77	62	8.7
Catalase	7.4	0	0	5.6	8.8	0

[a] The activities of the enzymes were determined as described in the legend to Fig. 4;
[b] The concentration of CAP (chloramphenicol) and CHI (cycloheximide) were the same as those shown in Table 5. Incubation was carried out for 6 h with or without PIN (pyridoxine, 100 μg ml^{-1}); Details are given in Ref. [52]

enzymes in this process was confirmed by the inhibitory effects of chloramphenicol and cycloheximide on the increases in the enzyme activities. Pyridoxine caused marked increases in the activities of not only respiratory enzymes but also other heme-containing enzymes, lactate dehydrogenase and catalase. The developments of such enzymes were also inhibited by cycloheximide. In contrast to the case of the respiratory enzymes, chloramphenicol did not affect the increments of the enzyme activities, being consistent with the fact that the syntheses of these enzymes occur in the cytoplasmic system of protein synthesis [40,66]. These results strongly suggest that the levels of the heme-containing enzymes can be determined by the content of hemes in the yeast cells as in the case of hemoglobin biosynthesis [24]. An alternative explanation might be possible: the apoproteins of the heme-containing enzymes would be present even in the thiamine-grown cells and certain enzymes participating in heme biosynthesis would be newly synthesized during the incubation of the cells with pyridoxine.

The respiratory adaptation with pyridoxine was accompanied by the appearance of cytochromes. The peak of cytochrome c (552 nm) was lower than that of cytochrome b (563 nm) in contrast to the case of growing culture in the presence of added pyridoxine. Similar phenomenon has been reported in the cells of a heme-lacking mutant of *S. cerevisiae* grown with added ALA [23] and in the respiratory adapted cells of *S. cerevisiae* [5].

Table 7. Glucose effect on the development of the respiratory activity caused by pyridoxine

Conditions	Cytochrome oxidase (nmol min^{-1} mg^{-1} protein)
Before incubation	0.2
After incubation[a] with	
None	3.9
Glucose[b]	1.7
Pyridoxine	20
Pyridoxine plus glycose[b]	2.0

[a] Incubation with or without added pyridoxine (100 µg ml^{-1}) was carried out for 4 h;
[b] Glucose was added at the initiation of the incubation at a final concentration of 4%;
Conditions were the same as those described in Table 5

Pyridoxine-dependent development of respiratory activity was under the influence of glucose (Table 7) as in the usual "respiratory adaptation" in which anaerobically grown cells develop their respiratory system during aerobic incubation [7, 30]. A high concentration of glucose (4%) inhibited the pyridoxine-dependent development of cytochrome oxidase activity almost completely.

Small but appreciable increases in the respiration rate and the activities of respiratory enzymes were observed even in the absence of pyridoxine during incubation of the thiamine-grown cells (cf. Fig. 9, Tables 5, 6). However, such pyridoxine-independent increments were much less than those brought about by pyridoxine except for the case of succinate-cytochrome c oxidoreductase. The pyridoxine-independent increases in activity occurred only in the respiratory enzymes and not in lactate dehydrogenase and catalase. This can be explained by the release of the respiratory enzymes from the glucose effect, because the cells grown in the medium containing 5% glucose were washed and incubated in the adaptation medium containing a low concentration of glucose (0.1%). In fact, the increasé in cytochrome oxidase activity was inhibited by glucose at a high concentration (4%) (cf. Table 7). The syntheses of lactate dehydrogenase and catalase might be less sensitive to glucose repression than those of the respiratory enzymes. Even the respiratory enzymes are apparently resistant to the glucose effect in a growing culture, as described above; the addition of pyridoxine to the growth medium containing thiamine permitted the cells to grow normally with normal levels of respiratory enzymes even in the presence of more than 4% of glucose. These observations suggest that the extent of the glucose effect depends not only on the characteristics and localization of the enzymes but also on the biological state of cells.

2.4.2 Effect of δ-Aminolevulinate [55]

ALA also caused respiratory adaptation of the thiamin-grown cells. In the same resting cell system as described above, ALA (at a final concentration of 1 mM) exhibited a similar effect to that of pyridoxine, resulting in a significant increase in the activities of the heme-containing enzymes (Table 8). After the incubation with ALA, the

Table 8. Increase in the activities of respiratory enzymes and other hemo-containing enzymes during incubation of the thiamine-grown cells with ALA

Enzyme	Specific activity (nmōl min^{-1} mg^{-1} protein)					
	Control cells	Thiamine-grown cells				
		Before incubation	After incubation with			
			None	ALA	ALA plus CAP	ALA plus CHI
Succinate-cytochrome c oxidoreductase	3.6	0.16	0	2.8	0.31	0.39
NADH-cytochrome c oxidoreductase	28	1.0	1.5	12	1.9	1.6
Cytochrome oxidase	35	0.8	1.7	8.6	1.7	1.5
NADH oxidase	29	3.5	1.9	15	1.0	2.2
Lactate dehydrogenase	36	0	0	33	7.4	0
Catalase	7.4	0	0	27	18	0

Conditions were the same as those in Table 6. Details are described in Ref. [55]

activities of all of these enzymes except catalase were markedly increased but somewhat lower than the activities in the control cells. Incubation of the thiamine-grown cells in the absence of added ALA caused practically no change in the activities of the enzymes. The increase in the activities of the respiratory enzymes by ALA was inhibited almost completely upon addition of either cycloheximide or chloramphenicol. The occurrence of the other heme-containing enzymes was blocked completely by cycloheximide but only partially by chloramphenicol. The effects of these inhibitors were similar to those observed in the case of the respiratory adaptation with pyridoxine (cf. Table 6). As discussed above, the heme-containing enzymes other than the respiratory enzymes, especially catalase, may be synthesized only on cytoplasmic ribosomes. The effect of the antibiotics indicated that de novo protein syntheses were involved in the ALA-induced enhancement of the enzyme activity.

The occurrence of the respiratory chain enzymes in the thiamine-grown cells upon incubation with ALA was confirmed by taking cytochrome spectra of the cells before and after the incubation. The cells incubated with ALA for 5 h showed the peaks corresponding to cytochrome aa_3, b, and c. Different from the case of respiratory adaptation caused by pyridoxine, ALA gave a normal pattern of cytochrome spectra, that is, the peak of cytochrome c was higher than that of cytochrome b. The addition of porphobilinogen (200 $\mu g\ ml^{-1}$) or hemin (100 $\mu g\ ml^{-1}$) was also found to cause the occurrence of cytochrome oxidase, a typical heme-containing enzyme, during incubation of the thiamine-grown cells, although the effects of these substances were smaller than that of ALA. The enzyme activities, expressed in nanomoles per minute per mg of protein, were as follows: before incubation, 0.76; after incubation 4.6 (without supplement), 23.9 (with ALA), 13.8 (with porphobilinogen), and 13.0 (with hemin). The increase in activity during incubation without heme-related compounds may be due to the release from glucose repression, as discussed previously.

The effect of ALA addition on the cellular content of heme precursors, coproporphyrin III and protoporphyrin IX, was examined in the resting cell system. The thiamine-grown cells had no appreciable amounts of the porphyrins, as mentioned above (cf. Table 2). Incubation of the cells for 5 h without ALA caused a small but significant increase in the content of the porphyrins (Table 9). However, the addition of ALA brought about a much larger increase in the porphyrin contents.

Table 9. Increase in the cellular content of porphyrins during incubation of the thiamine-grown cells with ALA

Porphyrin	Porphyrin content (nmol g^{-1})		
	Before incubation	After incubation with	
		None	ALA
Coproporphyrin III	0	0.004	0.023
Protoporphyrin IX	0.039	0.12	0.38

Conditions for incubation of the thiamine-grown cells were the same as those in Table 6. Porphyrins were determined as described in Table 2. Details are shown in Ref. [55]

The respiratory adaptation by pyridoxine and by ALA in the thiamine-grown cells would provide further evidence for the fact that the supply of ALA is blocked in the cells owing to the thiamine-induced vitamin B_6 deficiency. In addition, this respiratory adaptation system would serve as a unique experimental means for the study of mitochondrial biogenesis. We have found that mitochondria from the thiamine-grown cells have well-developed cristae [54] despite the absence of cytochromes and heme-containing enzymes. Furthermore, the mitochondria possess the normal activity of oligomycin-sensitive ATPase. However, the lipid composition was found to be altered significantly. Cardiolipin, a marker of mitochondrial membrane, was also detected but in a reduced amount. The content of other phospholipids was also changed. Unsaturated fatty acid content was reduced and, on the other hand, short-chain fatty acid content was increased as observed in the whole cells (described below). One of the most striking characteristics of the mitochondria of the thiamine-grown cells is the altered lipid composition as compared with those of anaerobically grown yeast cells or heme deficient mutant cells which are also deficient in respiratory activity but contain significant levels of unsaturated fatty acids and ergosterol owing to the supplementation of these lipids to support growth. It would be of particular interest to investigate the change in the composition of mitochondrial membrane during the respiratory adaptation caused by pyridoxine or ALA and its relation to the mitochondrial functions.

3 Accumulation of Thiamine in Yeast Cells and Its Relation to the Thiamine-Induced Respiratory Deficiency [56]

To know whether the effects of thiamine and pyridoxine occur universally in yeasts by the same mechanism, several yeasts other than *S. carlsbergensis* were examined for the sensitivity to thiamine in respect of growth, vitamin B_6 content, and cytochrome oxidase activity.

It should be noted that the remarkable effect of thiamine on the growth of *S. carlsbergensis* 4228 is observed at low concentrations not exceeding the physiological demands (less than 1 ng of thiamine · HCl per ml of medium) [71]. Thiamine is known to be incorporated in large amounts into the cells of *S. cerevisiae* [32,82]. The accumulation of thiamine was thought to account for the above-mentioned pronounced effects of the vitamin.

As will be described in this chapter, growth depression by thiamine occurred not only in *S. carlsbergensis* 4228 but in the other *Saccharomyces* yeasts tested and was invariably accompanied by the marked decrease in cellular vitamin B_6 content and in cytochrome oxidase activity. In all the thiamine-sensitive yeasts, the effects of pyridoxine and ALA were quite similar to those found in *S. carlsbergensis* 4228, suggesting that the inhibitory action of thiamine proceeds by the same mechanism among these yeasts through the vitamin B_6 deficiency. Yeasts belonging to other genera were not influenced by thiamine. Further, in only the cells of the sensitive yeasts, thiamine was accumulated abundantly in the non-esterified form. This fact indicated that the accumulation of non-esterified thiamine would be the primary event for the thiamine effect in *Saccharomyces* yeasts.

Eight yeast species belonging to *Saccharomyces*, *Kluyveromyces*, *Schizosaccharomyces*, and *Candida* were examined for growth response to thiamine and pyridoxine under the same conditions employed to cultivate *S. carlsbergensis* 4228 cells. Similar to *S. carlsbergensis* 4228, the growth of *S. cerevisiae*, *S. sake*, and *S. oviformis* was also inhibited by adding thiamine to the vitamin B_6-free medium (Fig. 10). Thiamine caused an elongation of the lag phase of growth, a lowering of the growth rate and of the maximum growth level. These thiamine-sensitive species (all of them belonged to *Saccharomyces*) were termed Type I yeasts. The growth inhibition caused by thiamine was completely abolished by concomitant addition of pyridoxine. On the other hand, ALA did not bring about the normal growth profile; the prolonged lag phase was not improved although the rate and maximum level of growth were considerably increased in all Type I yeasts. The growth of the other four yeasts was not inhibited by thiamine (termed Type II yeasts).

As shown in Table 10, cellular vitamin B_6 content of Type I yeasts was high during early growth (the mid-exponential phase) and decreased rapidly in the control cells. The thiamine-grown cells of Type I yeasts contained markedly low amounts of vitamin B_6 throughout the growth period. The difference in vitamin B_6 content between the control and the thiamin-grown cells was most marked at the mid-exponential phase of growth. In Type II yeasts, thiamine exerted little effect on cellular vitamin B_6 content during early growth. These results suggest a close relationship between the decrease in the vitamin B_6 content at the early growth phase and the growth depression.

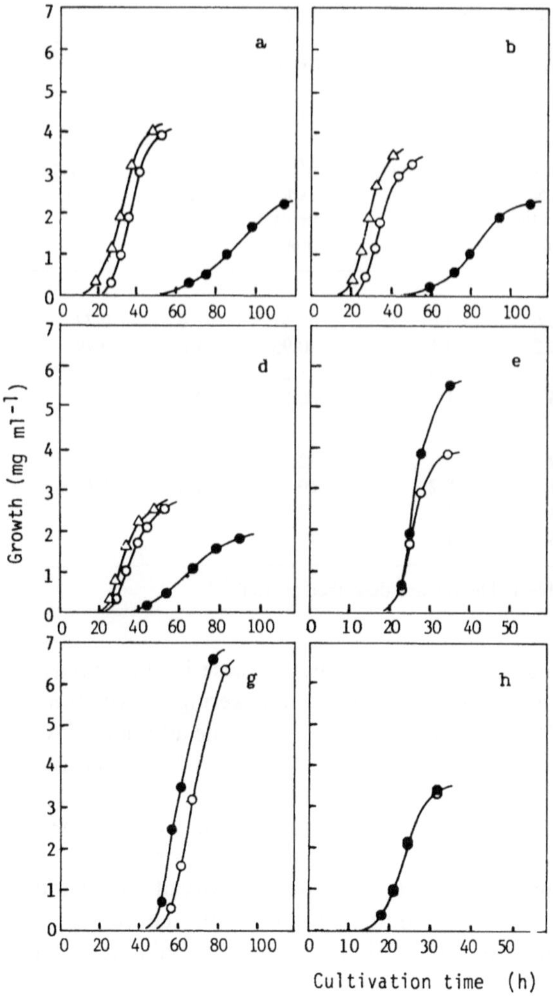

Fig. 10a—h. Effects of thiamine and pyridoxine on the growth of various yeasts [56]. Thiamine · HCl and pyridoxine · HCl were added to the culture medium at final concentrations of 1 μg and 0.02 μg per ml, respectively. Culture conditions were the same as those in Fig. 1. **a** *S. carlsbergensis* 4228 [ATCC 9080]; **b** *S. cerevisiae* [ATCC 7753]; **c** *S. sake* Kyokai No. 7; **d** *S. oviformis* [IFO 0262]; **e** *K. fragilis* [IFO 0288]; **f** *K. lactis* [IFO 1090]; **g** *Sch. pombe* [IFO 0346] and **h** *C. utilis* [IFO 1086]. (○) Control cells, (●) thiamine-grown cells, (△) thiamine and pyridoxine-grown cells

Cytochrome oxidase activity in the control cells of the yeasts, except for *Sch. pombe* and *C. utilis*, increased with growth (Table 11). In all Type I yeasts, cytochrome oxidase activity was markedly low when the cells were grown with thiamine, as observed in *S. carlsbergensis* 4228 [51–53, 55] (cf. Fig. 4). The lowered level of the activity was maintained during the cultivation. The activity of *S. oviformis* cells was exceptionally increased at the stationary phase. However, the increased activity was still much lower than the activity in the control cells. The thiamine effect was prevented by the

Table 10. Effect of thiamine on the cellular vitamin B_6 content in yeasts

Yeast	Vitamin B_6 content [ng pyridoxine · HCl mg^{-1}]					
	Control cells			Thiamine-grown cells		
	Mid-exponential phase	Late exponential phase	Stationary phase	Mid-exponential phase	Late exponential phase	Stationary phase
Type I						
S. carlsbergensis	12	6.6	2.6	0.61	1.6	0.77
S. cerevisiae	7.0	5.2	3.1	0.95	1.0	0.79
S. sake	27	11	2.3	1.2	0.89	0.71
S. oviformis	80	46	21	6.0	2.7	1.3
Type II						
K. fragilis	19	19	20	22	17	5.4
K. lactis	11	11	5.2	10	4.3	3.1
Sch. pombe	12	24	13	13	22	6.2
C. utilis	9.2	8.0	11	10	8.7	5.9

Conditions were similar to those in Table 1. Details are described in Ref. [56]

concomitant addition of pyridoxine, indicating that the decrease in the activity of cytochrome oxidase resulted from the thiamine-induced lowering of vitamin B_6 content as described in the case of *S. carlsbergensis* 4228 [53,55]. The addition of ALA also caused the prevention of the thiamine effect as observed in *S. carlsbergensis* 4228 [55] (cf. Fig. 7).

The oxidase activity level in Type II yeasts was not affected by thiamine. In *Sch. pombe*, a fairly significant decrease (one half of the control cells) was observed at the mid exponential phase. However, the oxidase activity of the thiamine-grown cells of *Sch. pombe* was still high compared even with the control cell activities of the other yeasts. Also, the rate of thiamine-induced reduction of the activity in this yeast was much smaller than those in Type I yeasts.

Cells of *S. cerevisiae* accumulate added thiamine in large amounts [32,82] and the transported thiamine has been found predominantly in the non-esterified form [32]. Thiamine uptake by the various yeast cells was determined by a modified method of Iwashima et al. [32]. Cells harvested at the early exponential phase were washed and suspended in 50 mM potassium phosphate buffer (pH 5.0) containing 0.1 mM glucose at a final concentration of 100 µg of dry cells per ml. The cell suspension (5 ml) was incubated at 30 °C for 15 min with gentle shaking. Thiamine uptake was initiated by adding 0.3 µmol (0.28 µCi) of [^{14}C]thiamine (thiazole-2-^{14}C-thiamine · HCl, 14.0 Ci mol^{-1}). Incubation was continued for 60 min with constant shaking and the uptake of radioactivity into the cells was measured at 0, 10, 30 and 60 min after the addition of [^{14}C]thiamine. At each time, 1 ml portions of the reaction mixture were withdrawn and the cells were collected on a membrane filter followed by washing once with the above buffer. After drying, the radioactivity was measured.

The time course of [^{14}C]thiamine uptake by the cells was followed for each yeast. The radioactivity was transported into the cells in both yeast types without appreciable

Table 11. Effects of thiamine, pyridoxine and ALA on the activity of cytochrome oxidase in yeasts [56)]

Yeast	Cytochrome oxidase specific activity (nmol min^{-1} mg^{-1} protein)[a]											
	Control cells			Thiamine-grown cells			Thiamine and pyridoxine-grown cells			Thiamine and ALA-grown cells		
	Mid-exponential phase	Late exponential phase	Stationary phase	Mid-exponential phase	Late exponential phase	Stationary phase	Mid-exponential phase	Late exponential phase	Stationary phase	Mid-exponential phase	Late exponential phase	Stationary phase
Type I												
S. carlsbergensis	21	24	42	0.81	0.54	0.30	18	24	25	29	19	12
S. cerevisiae	22	39	52	0.43	0.61	0.25	14	13	43	19	20	16
S. sake	21	37	68	0.79	0.78	0.24	19	19	44	17	17	9.1
S. oviformis	18	23	52	0.28	0.32	3.6	18	18	57	15	15	46
Type II												
K. fragilis	89	105	212	82	108	118	—	—	—	—	—	—
K. lactis	21	25	41	24	26	37	—	—	—	—	—	—
Sch. pombe	230	100	93	118	88	127	—	—	—	—	—	—
C. utilis	90	71	67	92	67	63	—	—	—	—	—	—

—, Not determined;
[a] Cytochrome oxidase was assayed as described in the legend to Fig. 4

Table 12. Thiamine accumulation in yeasts

Yeast	[^{14}C]Thiamine accumulated (nmol mg^{-1})
Type I	
S. carlsbergensis	400
S. cerevisiae	259
S. sake	224
S. oviformis	225
Type II	
K. fragilis	6.6
K. lactis	3.3
Sch. pombe	9.4
C. utilis	3.6

Cell suspensions (100 µg dry cells per ml) were prepared and incubated with [^{14}C]thiamine as described in the text except that samples were withdrawn from the cell suspensions during the incubation at appropriate time intervals. The radioactivity of the cells on the membrane filter was counted. The time course of the radioactivity transported into the cells was plotted. Accumulation of [^{14}C]thiamine in the cells was calculated from the maximum levels attained. See Ref. [56] for details

lag. As judged by paper chromatography of the extracts of the radioactive cells, thiamine transported into Type I yeasts was recovered only in the non-esterified form. On the other hand, significant levels of radioactivity were detected on the spots corresponding to thiamine pyrophosphate and thiamine monophosphate in the case of Type II yeasts. The levels of thiamine transported in 60 min incubation are shown in Table 12. The results clearly indicated that the accumulation in cells of Type I yeasts was much higher than that in Type II yeasts (more than 24-fold). The possibility that thiamine had become attached to the cell surface of Type I yeasts by some ionic interaction was excluded since washing of the cells with 0.1 M HCl or 0.3 M KCl caused no appreciable reduction of the radioactivity. The intracellular concentration of accumulated thiamine in *S. carlsbergensis* was calculated from the result in Table 12 to be 190 mM assuming that the cellular water space in the yeast cells is 2.1 µl per mg dry weight [63]. This concentration is about 9500 times that of the exogenously remaining thiamine. A similar result was obtained by Iwashima et al. [32] in *S. cerevisiae*. Such a level of thiamine accumulation in Type I yeasts in extraordinarily high compared with that in *E. coli* [37,38] or in other bacteria [28,61].

Thiamine has no inhibitory effect on the growth of *S. carlsbergensis* when added after the initiation of growth [6]. This fact strongly suggested that thiamine exerts its effect at extremely early stages of growth. Hence, the cellular content of thiamine of Type I and Type II yeasts at such a growth phase was determined in the thiamine-grown culture. To obtain sufficient amounts of cells for measuring thiamine content during the lag phase of growth, the inoculum size was increased by 40 times (20 µg dry cells per ml medium). In Type I yeasts, the free thiamine content began to increase immediately after cultivation had started and reached maximum levels before appreciable proliferation of cells occurred (Fig. 11), reflecting the high capacity of

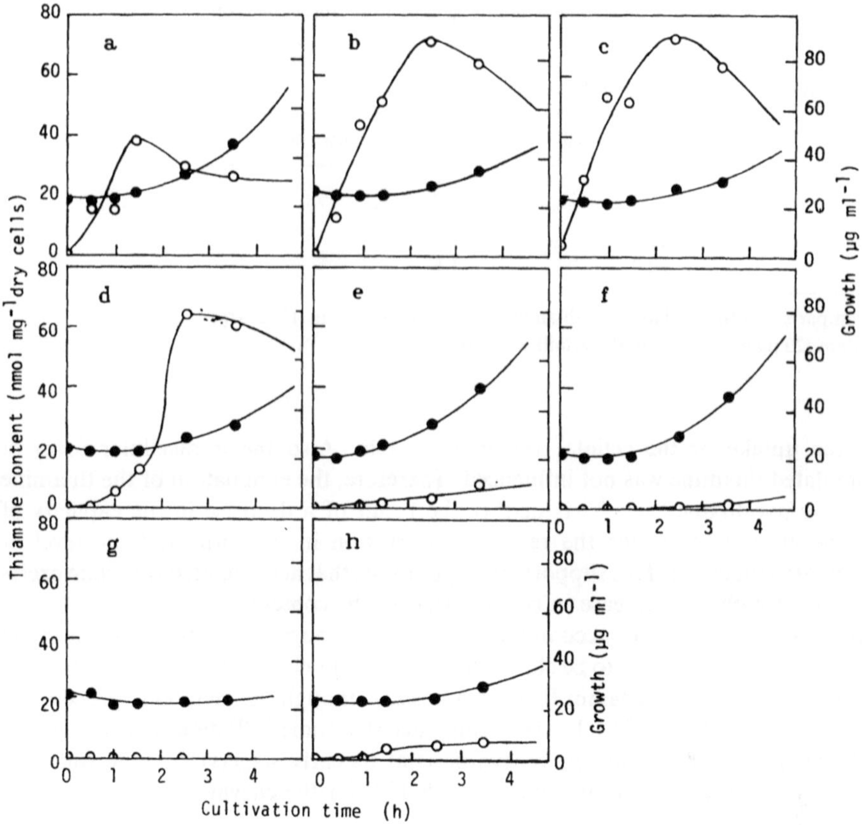

Fig. 11a—h. Accumulation of thiamine in growing yeast cells at an extremely early growth stage [56]. Cells harvested at the stationary phase were inoculated at a final concentration of 20 μg dry cells per ml of the medium containing thiamine (1 μg per ml) and cultivated. Cells were harvested at the indicated time of cultivation, and free thiamine was extracted from the cells and determined by a thiochrome method (see Ref. [56] for details). **a** *S. carlsbergensis*; **b** *S. cerevisiae*; **c** *S. sake*; **d** *S. oviformis*; **e** *K. fragilis*; **f** *K. lactis*; **g** *Sch. pombe*, **h** *C. utilis*. (○) Accumulation of free thiamine, (●) growth

thiamine accumulation of the cells. In contrast, Type II yeasts showed a much lower content of free thiamine in this short period of cultivation.

As described earlier, the thiamine-grown cells of *S. carlsbergensis* 4228 contained lower amounts of vitamin B_6 than the control cells throughout the cultivation period (cf. Table 1,10). The time course of vitamin B_6 content at the extremely early stage of growth was determined using the same yeast (Table 13). For comparison with the changes in thiamine content of the thiamine-grown cells shown in Fig. 11, the large inoculum was used. These results indicated that the rapidly accumulated thiamine (cf. Fig. 11-A) inhibited the increase in the content of vitamin B_6 which was detected in the control cells.

Addition of pyridoxine to the medium used for thiamine uptake (2 μM) or to the thiamine-supplemented medium for growth (100 nM) had no effect on the amount of

Table 13. Effect of thiamine on cellular vitamin B_6 content in *Saccharomyces carlsbergensis*

Cultivation time (h)	Vitamin B_6 content (ng mg^{-1})	
	Control cells	Thiamine-grown cells
0	2.2	2.2
4	4.6	0.93
8	11	2.0
12	2.0	1.6

For comparison with the changes in thiamine content shown in Fig. 11, a large inoculum [20 µg ml^{-1}] was used. Details are shown in Ref. [56]

thiamine uptake or the cellular thiamine content. Also the intracellular form of accumulated thiamine was not influenced. Therefore, the elimination of the thiamine effects by pyridoxine [51–53, 55] is neither the result of a decrease in the capacity of thiamine accumulation nor the results of alteration in the form and the level of accumulated thiamine. This supports the hypothesis that accumulated thiamine exerts its effect through the decrease in cellular vitamin B_6 content.

The reason why thiamine accumulated in the cells brought about the decrease in the vitamin B_6 content remains to be elucidated. It is very likely that thiamine affects some step(s) in the process of vitamin B_6 biosynthesis. Although the metabolic process has not been elucidated [70, 77, 89], it has been suggested that some TPP-dependent enzyme(s) would participate in vitamin B_6 biosynthesis at its early stage(s) [68]. Free thiamine accumulated in the cells would compete with TPP for the enzyme(s).

4 Effects of Thiamine and Pyridoxine on Lipid Metabolism in *Saccharomyces carlsbergensis* [50, 58–60]

The preceding chapters described that thiamine caused cytochrome deficiency through vitamin B_6 deficiency in the cells of *Saccharomyces* yeasts. Lines of evidence have been accumulated for the participation of microsomal cytochromes in lipid metabolism. It was therefore of interest to investigate the lipid metabolism and composition in the thiamine-grown cells which are deficient in cytochromes.

4.1 Alteration in Fatty Acid Composition [58]

Figure 12 shows the results of gas-liquid chromatographic analysis of fatty acid methyl esters in the cells of *S. carlsbergensis* 4228 grown to the mid-exponential phase. The control cells contained mainly palmitoleic and oleic acids and minor amounts of the corresponding saturated fatty acids, palmitic and stearic acids. A striking decrease was observed in the content of unsaturated fatty acids of the thiamine-grown cells, whereas that of saturated fatty acids was slightly increased. Cells growing with pyridoxine exhibited the normal composition of fatty acids even in the presence of thiamine (the thiamine and pyridoxine-grown cells). The ratio of

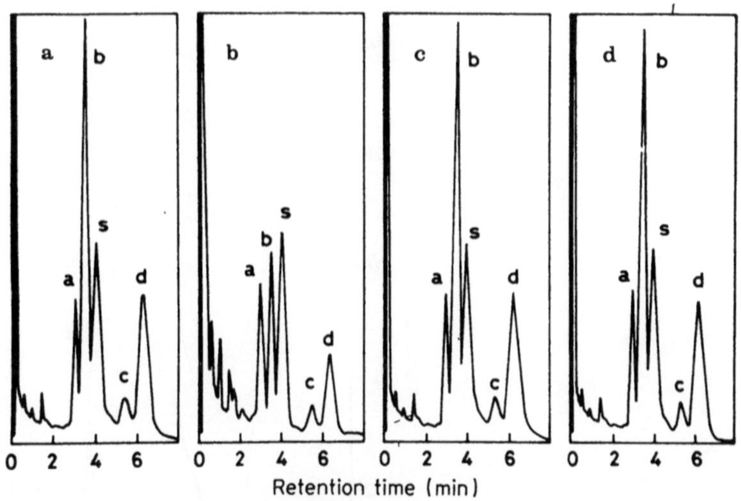

Fig. 12a—d. Gas-liquid chromatogram of fatty acid methyl esters from the control cells **a**; the thiamine-grown cells **b**; the thiamine and pyridoxinegrown cells **c**; and cells grown with pyridoxine alone **d**. [58] Individual methyl esters were separated as described in Ref. [58], and marked as follows: (a) palmitate, (b) palmitoleate, (c) stearate, (d) oleate, and (s) internal standard (heptadecanoate)

monounsaturated to saturated fatty acids was about 9:1 in the control cells, whereas it decreased to about 2:1 in the thiamine-grown cells.

The cytochrome deficiency in the thiamine-grown cells was considered to induce the decrease in the unsaturated fatty acid content, since the desaturation of long chain fatty acids is known to occur by the reaction catalyzed by cytochrome b_5 in rat liver [65] and yeast [83].

4.2 Alteration in Sterol Composition [50, 59, 60]

Nonsaponifiable lipids extracted from the cells at the mid-exponential growth phase were analyzed by gas-liquid chromatography. As revealed in Fig. 13, the control cells showed a normal composition of sterol. The cells contained mainly ergosterol and zymosterol, and minor amounts of their precursors, squalene and lanosterol. In contrast, the thiamine-grown cells exhibited a quite different pattern of sterol composition. Zymosterol and ergosterol were almost completely absent in the cells, and the amounts of squalene and lanosterol were higher than those of the control cells. The presence of unidentified sterols was detected.

It is known that ergosterol is the predominant sterol in yeasts and molds, comprising generally 90% of the total sterols [25, 73]. Moreover, the sterol is considered to be the final product of sterol biosynthesis [14]. The unidentified sterols found in the thiamine-grown cells would be intermediary metabolites in sterol biosynthetic pathways. Hence, we investigated in detail the composition of "the thiamine-grown cells" and determined the structure of the unidentified sterol. From the sterol composition the altered pathways of sterol biosynthesis in the thiamine-grown cells were proposed.

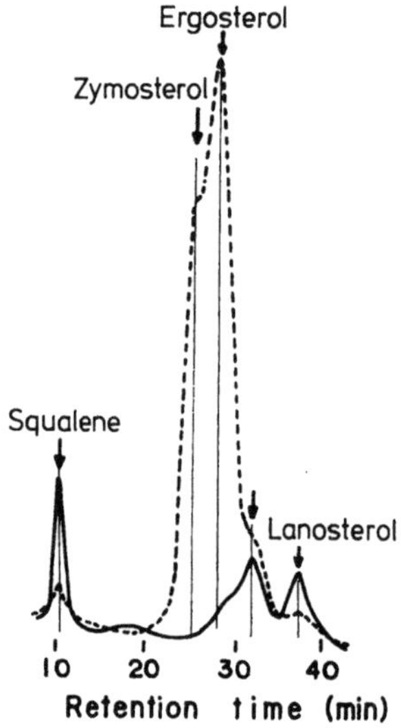

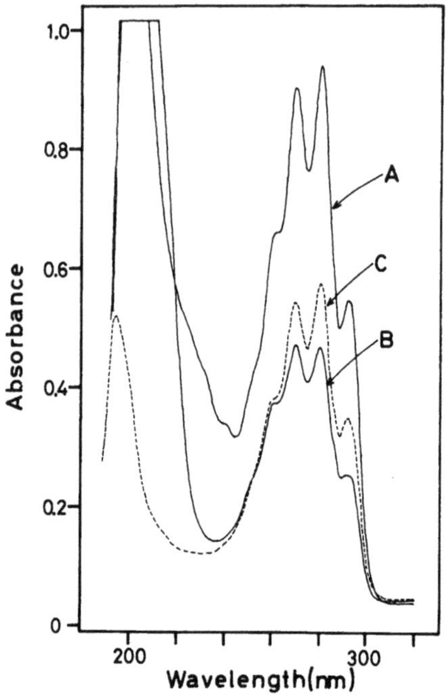

Fig. 13. Gas-liquid chromatograms of nonsaponifiable lipids from *S. carlsbergensis* cells harvested at the mid-exponential growth phase [50]. Samples were prepared as described in Ref. [50]. Comparable amounts of the samples of the control cells (– – – –) and the thiamine-grown cells (———) were injected. The gas-liquid chromatography was performed on 1.5% SE-30

Fig. 14. Ultraviolet absorption spectra of nonsaponifiable lipid extracts from the control cells (A) and the thiamine-grown cells (B) [60]. Nonsaponifiable lipids were extracted from the cells with n-hexane and analyzed spectrophotometrically. (C) Authentic ergosterol (0.02 mg per ml)

The absorption spectrum of nonsaponifiable lipid extracts from the control cells coincided completely with that of an authentic sample of ergosterol (Fig. 14). This result together with the data in Fig. 13, indicated the presence of ergosterol in the cells. The content of ergosterol was calculated as 5.2 mg per g dry cells from the intensity of absorbance at 282 nm. The extracts from the thiamine-grown cells also gave a representative absorption spectrum showing the presence of the $\Delta^{5,7}$-double bond in the sterol structure, in spite of the fact that the cells had no ergosterol as judged by gas-liquid chromatography (cf. Fig. 13). The content of the $\Delta^{5,7}$-sterol(s) was calculated as 2.0 mg per g dry cells in the same way as above. It is known that the introduction of unsaturation at C-5 occurs at the late stage of ergosterol formation in yeasts [14]. Thus the $\Delta^{5,7}$-sterol(s) would be the main sterol component(s) of the cellular membranes of the thiamine-grown cells and probably plays an important role in membrane functions in place of ergosterol.

As shown in Fig. 15a, hydrogenated extracts from the control cells exhibited two major peaks of sterol on the gas-liquid chromatogram. The retention times of these peaks agreed with those of authentic samples of cholestan-3β-ol and ergostan-3β-ol, respectively, indicating that C_{27}- and C_{28}-sterols are predominant in the control cells. The C_{27}- and C_{28}-sterols were considered to be zymosterol and ergosterol, respectively, as judged by the gas-liquid chromatogram obtained previously (cf. Fig. 13). The cellular contents of zymosterol and ergosterol were calculated from the peak areas to be 3.1 and 5.3 mg per g dry cells, respectively. The value of ergosterol was in fair agreement with that obtained from the intensity of ultraviolet absorption. In addition, the content of total sterol was estimated to be 10.5 mg per g dry cells by gas-liquid chromatography. Then the contents of zymosterol and ergosterol were calculated as 30 and 51 % of total sterol, respectively. These values are comparable to those reported for other yeasts such as *Saccharomyces cerevisiae* [15] and *Candida utlilis* [16].

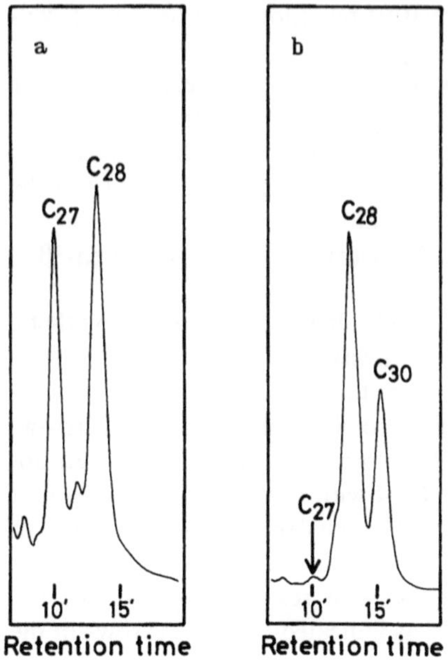

Fig. 15a and b. Gas-liquid chromatograms of hydrogenated nonsaponifiable lipid extracts from the control cells **a** and the thiamine-grown cells **b** [60]. Nonsaponifiable lipid extracts were hydrogenated over PtO_2. The gas-liquid chromatography was performed on 3 % SE-30

In contrast, the presence of C_{27}-sterol was hardly detectable in the thiamine-grown cells (Fig. 15b). This was consistent with the result shown in Fig. 13 that zymosterol was absent in the cells. Instead, the cells were found to contain C_{30}-sterol as well as C_{28}-sterol. 24(25)-Dihydrolanosterol, in addition to ergostan-3β-ol, was used as the reference hydrogenated sterol in this case. The C_{30}-sterol may be lanosterol, and C_{28}-sterol would be, at least, one of the unidentified sterols described above.

The nonsaponifiable lipids from the control cells were separated into five bands on an $AgNO_3$-impregnated silica gel G plate (data not shown). Four of them had the same R_f values as those of authentic samples of squalene, ergosterol, zymosterol and lanosterol, respectively.

The nonsaponifiable lipid extracts from the thiamine-grown cells gave seven bands, among which only three bands showed the R_f values coinciding with those of known compounds. One of the bands having the same R_f value as that of ergosterol, was considered to correspond to one or more unidentified sterols described above. This band was shown by ultraviolet absorption spectrum to contain $\Delta^{5,7}$-sterols and only ergostanol was produced by hydrogenating the extract of this band over PtO_2 catalyst. The band gave one large peak and another small peak on a gas-liquid chromatogram. The relative retention time of the major component of this band agreed very closely with that of $\Delta^{5,7}$-ergostadien-3β-ol reported by Patterson [67]. The compound proved to be identical with an authentic sample of the sterol as judged by retention time and mass spectral fragmentation pattern. The cellular level of the sterol was determined to be 26.0% of total sterol. The minor component of the band was shown as ergosterol but the level was less than 1%. Occasionally, the band gave another sterol peak. The presence of $\Delta^{5,7}$-double bonds in this sterol was easily suggested by its behavior in the thin-layer chromatography. This sterol was considered to be $\Delta^{5,7,24(28)}$-ergostatrien-3β-ol by comparing its relative retention time to that of $\Delta^{5,7}$-ergostadien-3β-ol. There were three more bands separated on the $AgNO_3$-silica gel G plate. One of them was identified as $\Delta^{8,24(28)}$-ergostadien-3β-ol by its relative retention time and mass spectrum. There were no published data suggesting the structures of sterols corresponding to the other two bands. The extracts of the bands were then analyzed on the basis of the number of methyl groups at the C-4 position by using another $AgNO_3$-silica gel thin layer chromatography system. The sterols corresponding to the bands were assigned to 4α-methyl-$\Delta^{8,24(28)}$-ergostadien-3β-ol and 4α-methyl-$\Delta^{8,24(25)}$-cholestadien-3β-ol.

Fryberg et al. [14] presented evidence for a multiplicity of sterol biosynthetic pathways in yeast, and they reported major pathways of sterol biosynthesis in *S. cerevisiae* by using various nystatin-resistant mutants [15]. The major pathways for the enzymatic conversion of lanosterol to ergosterol are indicated by black arrows in Fig. 16. The metabolic steps are: (1) removal of the three extranuclear methyl groups of lanosterol at C-14 and C-4; (2) alkylation at C-24 with concomitant reduction of the double bond at C-25 and generation of a $\Delta^{24(28)}$-methylene; (3) isomerization of Δ^8 to Δ^7; (4) introduction of unsaturation at C-22; (5) reduction of the $\Delta^{24(28)}$-double bond; and (6) introduction of unsaturation at C-5. Our results suggested that the pathways are altered in the thiamine-grown cells as indicated by white arrows in the figure.

The accumulation of lanosterol suggests that the demethylation process is partially blocked in the thiamine-grown cells. The occurrence of 4α-methyl-$\Delta^{8,24(28)}$-ergostadien-3β-ol and the absence of zymosterol indicate that the alkylation at C-24 precedes the last nuclear demethylation step at C-4. $\Delta^{8,24(28)}$-Ergostadien-3β-ol may be formed mainly via 4α-methyl-$\Delta^{8,24(28)}$-ergostadien-3β-ol.

The presence of a large amount of $\Delta^{5,7}$-ergostadien-3β-ol clearly indicates that the isomerization of Δ^8 to Δ^7, the introduction of Δ^5, and the reduction of $\Delta^{24(28)}$ proceed freely in the thiamine-grown cells. From the normal pathways of sterol biosynthesis, one can consider that the Δ^8-Δ^7 isomerization occurs prior to the Δ^5 introduction. Significant amounts of $\Delta^{5,7,24(28)}$-ergostatrien-3β-ol were detected in some experiments. It is therefore reasonable to conclude that $\Delta^{5,7}$-ergostadien-3β-ol,

▶

Fig. 16. Sterol biosynthesis in yeast [60]. Major pathways operative are indicated by black arrows. Proposed pathways in the thiamine-grown cells are indicated by white arrows. Sterols detected in the thiamine-grown cells are indicated by symbol (*)

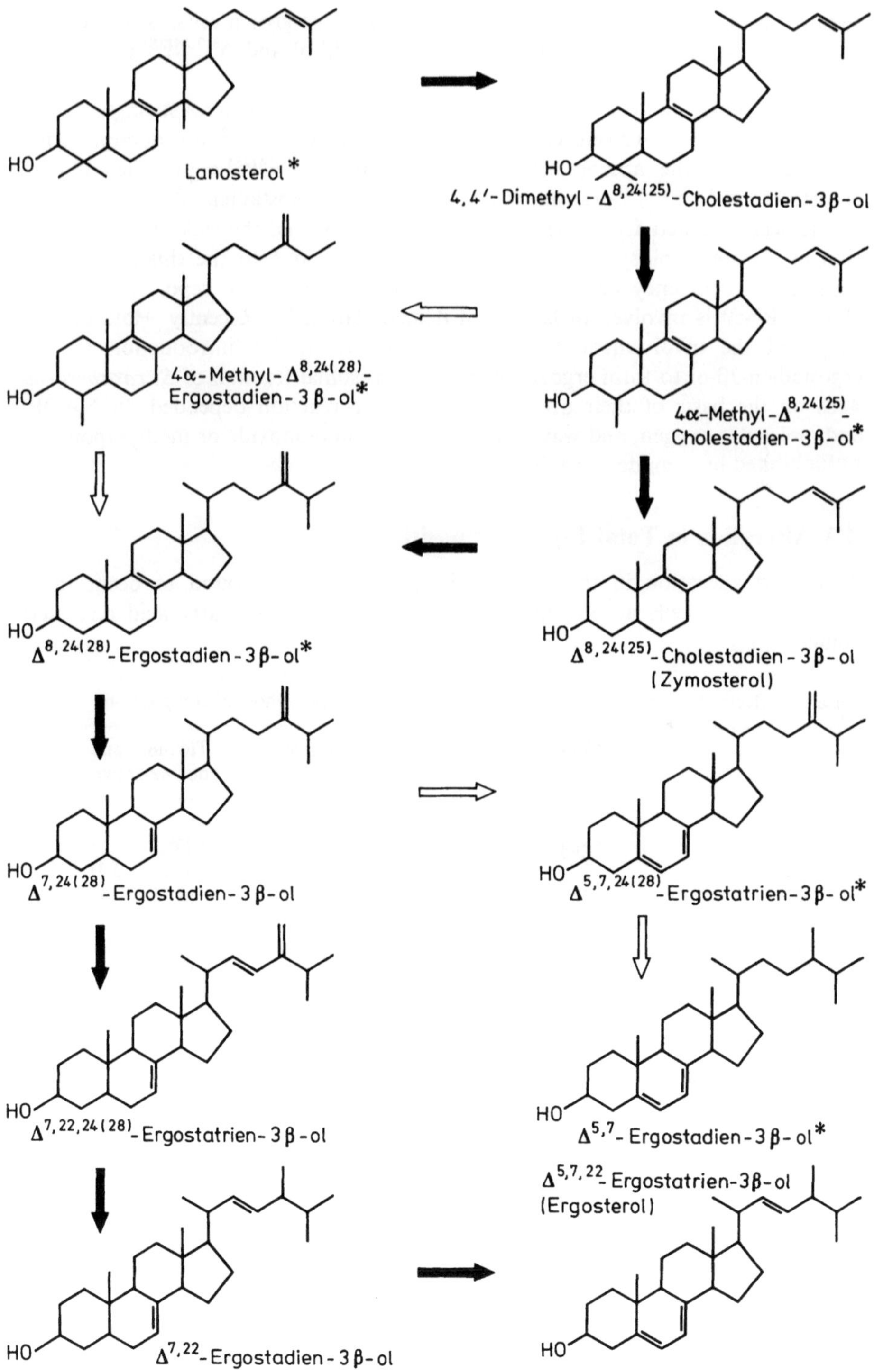
Lanosterol*
HO
4,4'-Dimethyl-Δ8,24(25)-Cholestadien-3β-ol
4α-Methyl-Δ8,24(28)-Ergostadien-3β-ol*
4α-Methyl-Δ8,24(25)-Cholestadien-3β-ol*
Δ8,24(28)-Ergostadien-3β-ol*
Δ8,24(25)-Cholestadien-3β-ol (Zymosterol)
Δ7,24(28)-Ergostadien-3β-ol
Δ5,7,24(28)-Ergostatrien-3β-ol*
Δ7,22,24(28)-Ergostatrien-3β-ol
Δ5,7-Ergostadien-3β-ol*
Δ5,7,22-Ergostatrien-3β-ol (Ergosterol)
Δ7,22-Ergostadien-3β-ol

the final product of sterol biosynthesis in the thiamine-grown cells, is formed from $\Delta^{8,24(28)}$-ergostadien-3β-ol via $\Delta^{7,24(28)}$-ergostadien-3β-ol and $\Delta^{5,7,24(28)}$-ergostatrien-3β-ol.

The accumulation of $\Delta^{5,7}$-ergostadien-3β-ol in place of ergosterol strongly suggests that the thiamine-grown cells completely lack the process of Δ^{22} introduction. This is consistent with the absence of the pathway from $\Delta^{7,24(28)}$-ergostadien-3β-ol to ergosterol via $\Delta^{7,22,24(28)}$-ergostatrien-3β-ol and $\Delta^{7,22}$-ergostadien-3β-ol.

The decreased activity of the demethylation process and the lack of Δ^{22} desaturation in the sterol biosynthesis were suggested to be due to the thiamine-induced cytochrome deficiency described above. In fact, it has been reported that cytochrome P-450 is involved in lanosterol demethylation [62]. Recently, Hata et al. [26] suggested the involvement of cytochrome P-450 in Δ^{22} introduction to $\Delta^{5,7}$-ergostadien-3β-ol to form ergosterol in the microsomal fraction of *S. carlsbergensis* 4228 on the basis of their observations that the reaction depended on NADPH and molecular oxygen, and was inhibited by carbon monoxide or methyrapone, but not inhibited by cyanide or azide.

4.3 Alteration in Total Lipid Composition [59]

A marked alteration in the total lipid composition was found to occur in the thiamine-grown cells accompanied by the alteration in the fatty acid and sterol compositions.

Table 14. Effects of thiamine and pyridoxine on the lipid and phospholipid compositions

Lipid	Control cells	Thiamine-grown cells	Thiamine and pyridoxine-grown cells
(A)			
Hydrocarbons	Squalene[a]	Squalene[a]	ND[b]
Free sterols	2.79 (3.0)[c]	1.31 (2.6)	4.68 (3.3)
Sterol esters	7.68 (8.4)	0.799 (1.6)	4.37 (3.0)
Diacylglycerols	4.53 (4.9)	7.67 (15.4)	10.6 (7.4)
Triacylglycerols	11.6 (12.6)	7.61 (15.3)	12.3 (8.5)
Phospholipids	25.8 (28.1)	18.7 (37.6)	39.3 (27.3)
Total lipids	91.9	49.7	144
(B)			
Phosphatidylinositol	0.13 (12.6)	0.19 (25.4)	0.10 (6.4)
Phosphatidylcholine plus phosphatidylserine	0.49 (47.6)	0.36 (48.1)	0.72 (45.9)
Phosphatidylethanolamine	0.36 (34.9)	0.20 (26.7)	0.70 (44.6)
Diphosphatidylglycerol	0.01 (1.0)	0.01 (1.3)	0.02 (1.3)
Total phospholipids[d]	1.03	0.75	1.57

Growth of the cells and extraction, separation and determination of lipids were carried out as described in Ref. [59];
(A) Lipid composition, mg g^{-1}; (B) phospholipid composition, mg lipid phosphorus per g dry cells;
[a] Squalene was identified, but not determined quantitatively;
[b] ND: not detected;
[c] Data in parentheses are % total lipids (A), and % total phospholipids (B)

Table 14-A shows that the content of various lipids was low in the thiamine-grown cells as compared with those in the control cells: 10% in sterol esters, 50% in free sterols, and 70% in either triacylglycerols or phospholipids. However, the amount of diacylglycerols was about 1.7 times higher than that of the control cells. The total lipid content was also markedly low in the thiamine-grown cells. Data in Table 14-A are also given as a percentage of total lipids in parentheses. The values for triacylglycerols and phospholipids were rather higher in the thiamine-grown cells. Thus, the thiamine-grown cells were considerably different from the control cells, not only in the absolute content but also in the composition of cellular lipids. These effects of thiamine were prevented by the addition of pyridoxine to the growth medium.

At least five different phospholipid spots were detected in the yeast cells by thin-layer chromatography (data not shown). Based on the specific color reactions and respective R_f values of individual lipid components, the major phospholipids of this yeast were tentatively identified as phosphatidylcholine (PC), phosphatidylethanolamine (PE), phosphatidylinositol (PI) and phosphatidylserine (PS). PC and PS could not be separated by the developing solvent used, but they were identified with Dragendorff reagent and ninhydrin, respectively. The spot of cardiolipin was too faint for its content to be discussed. (As mentioned in an earlier part of this paper, we are now investigating the structure and functions of the mitochondria from the thiamine-grown cells. Preliminary results with isolated mitochondria have shown that the content of cardiolipin is significantly low in the mitochondria from the thiamine-grown cells [54]).

It was found that the phospholipid composition of the thiamine-grown cells was qualitatively similar to that of the control cells. However, the addition of thiamine to the medium caused significant alteration in the level of individual phospholipids as shown in Table 14-B. The levels of PC plus PS and PE were substantially lower in the thiamine-grown cells. In contrast, the thiamine-grown cells contained a higher amount of PI. The increase in PI can be detected more clearly when the data are given as a percentage of total phospholipids (shown in parentheses). The phospholipid composition of the thiamine and pyridoxine-grown cells was similar to that of the

Table 15. Ratio of unsaturated to total fatty acids in neutral lipids and phospholipids

Lipids	Control cells (%)	Thiamine-grown cells (%)	Thiamine and pyridoxine-grown cells (%)
Sterol esters	86.2	34.6	87.7
Diacylglycerol	70.6	43.1	62.6
Triacylglycerol	86.5	57.5	74.2
Phosphatidylcholine plus phosphatidylserine	84.0	52.9	72.7
Phosphatidylethanolamine	78.2	62.1	76.1
Phosphatidylinositol	59.6	38.8	59.1

Growth of the cells and extraction of lipids were carried out as in Table 14. Lipids were separated by thin-layer chromatography and fatty acids from lipid classes were determined quantitatively by gas-liquid chromatography. For experimental details see Ref. [59]

control cells, although there was a significant difference in the amounts of individual phospholipids between these cells.

A marked difference in the ratio of unsaturated to total fatty acids was observed between the thiamine-grown cells and the control cells, that is, the ratio was 50% in the former and 80% in the latter. The level of short chain fatty acids increased with the decreased level of unsaturated fatty acids in the thiamine-grown cells.

Table 15 shows the composition of fatty acids present in major lipid classes; sterol esters, triacylglycerols, diacylglycerols, PC plus PS, PE, and PI. In every lipid class, especially in sterol ester, the content of unsaturated fatty acids was much lower in the thiamine-grown cells than in the control cells. The reduction in the amounts of unsaturated fatty acids in different lipid classes reflects the marked decrease in unsaturated fatty acid content in the thiamine-grown cells. The unsaturated fatty acid content of each lipid class in the thiamine and pyridoxine-grown cells was similar to that in the control cells, as was found for the cellular level of unsaturated fatty acids.

The effect of pyridoxine to abolish the thiamine effect on the lipid composition clearly demonstrates that the alteration in the composition of all lipid classes tested in the thiamine-grown cells was caused in general by the thiamine-induced vitamin B_6 deficiency. As discussed earlier, the vitamin B_6 deficiency can be considered to be the cause of cytochrome deficiency. It may be therefore reasonable that the alteration in the total lipid composition is the result of the decreases in the levels of unsaturated fatty acids and the loss of ergosterol, since the desaturation of fatty acids and the biosynthesis of ergosterol are catalyzed by microsomal cytochromes, as mentioned above. This concept would be at least partially supported by our finding that the unsaturated fatty acid content of individual lipids was decreased in a similar ratio (Table 15).

5 Effects of Thiamine-Induced Vitamin B_6 Deficiency and Resulting Respiratory Deficiency on Amino Acid Metabolism in *Saccharomyces carlsbergensis* [57)]

Vitamin B_6, in the form of pyridoxal phosphate and pyridoxamine phosphate, serves as a coenzyme for numerous enzymes which catalyze amino acid metabolism. Respiratory activity affects amino acid metabolism through TCA cycle. It is therefore reasonable to suppose that the metabolism is influenced by respiratory deficiency and/or vitamin B_6 deficiency caused by thiamine in *S. carlsbergensis* cells. Glutamate dehydrogenase (GDH), which occupies a strategic position in an important branch point between respiratory and amino acid metabolisms, is known to be metabolically regulated according to the environmental conditions such as glucose concentration, extent of aeration, and nitrogen source [8, 9, 49, 75, 84)]. NAD-linked GDH (NAD-GDH) catalyzes the dehydrogenation of glutamate to 2-oxoglutarate under aerobic conditions and is repressed by glucose and NH_4^+. NADP-linked GDH (NADP-GDH), in contrast, serves for the reduction of 2-oxoglutarate to glutamate in the presence of a relatively high concentration of glucose. Under the conditions affecting the levels of two GDH enzymes, their formation is regulated in a reciprocal manner. It has now

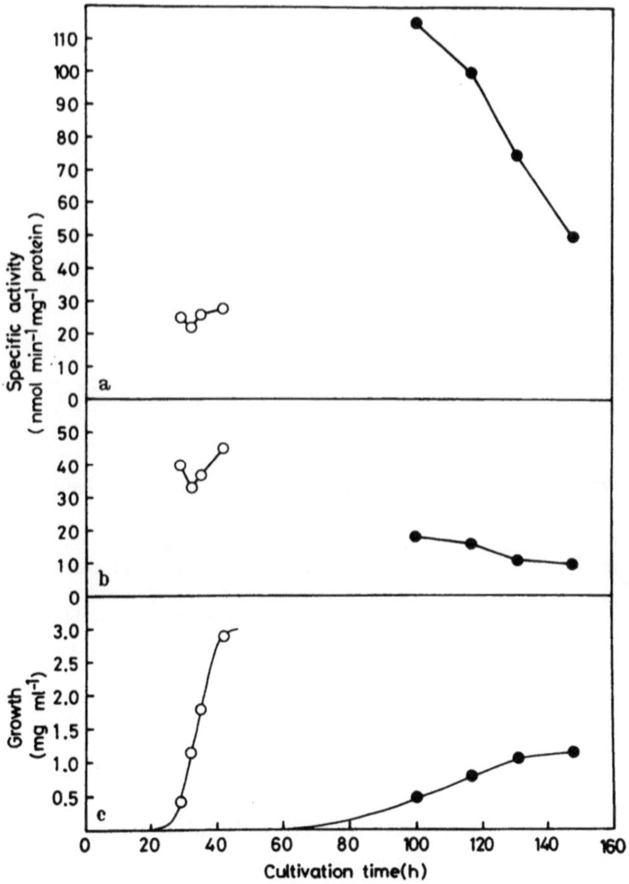

Fig. 17. Effect of thiamine on NAD-GDH and NADP-GHD activities [57]. Cells were harvested periodically, and the enzyme activities were determined in cell-free extracts as described in Ref. 57. (○) Without thiamine, (●) with thiamine. **a** NADP-GDH activity, **b** NAD-GDH activity, **c** growth

become one of the most interesting problems concerning GDH regulation whether any direct connection is present between the syntheses of these enzymes.

In this chapter, changes in the activities of the two GDH enzymes in the yeast cells growing under the influence of thiamine are described. Intracellular pools of amino acids and NH_4^+ were estimated in relation to GDH activities. Cellular activities of glutamate-oxalacetate transaminase (GOT) and glutamate-pyruvate transaminase (GPT) were also determined since these typical pyridoxal phosphate-linked enzymes are involved in governing the size of the glutamate pool and their activities may be influenced by the thiamine-induced vitamin B_6 deficiency.

Figure 17 shows the time course of NAD-GDH and NADP-GDH specific activities in cells growing aerobically with or without added thiamine. It should be noted that the medium contained a high concentration of glucose, 5%. A growth curve is also given for comparison. NAD-GDH activity was lower and NADP-GDH activity was markedly higher in the thiamine-grown cells as compared with those in the control

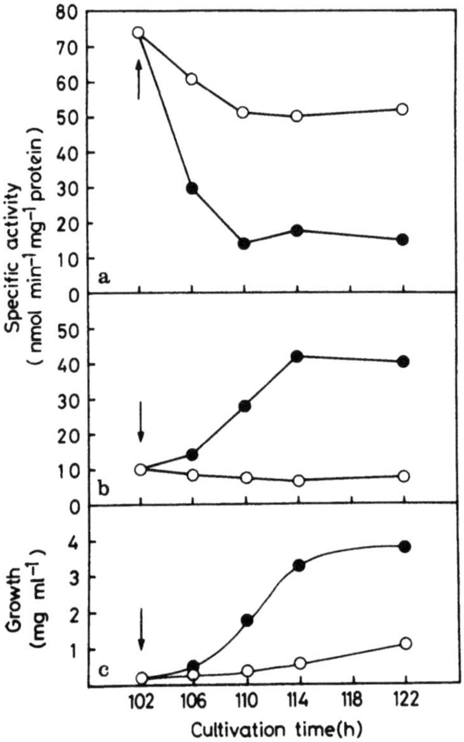

Fig. 18a—c. Effect of pyridoxine on GDH activities [57]. Pyridoxine was added at the time indicated by an arrow during cultivation. Cells were harvested periodically, and the GDH activities were measured as described in Fig. 17. (○) Without pyridoxine, (●) with pyridoxine. **a** NADP-GDH activity; **b** NAD-GDH activity; **c** growth

cells. Both vitamin B_6 content and respiratory activity were substantially lower in the thiamine-grown cells as described above. Addition of pyridoxine to the thiamine-supplemented culture at the mid-exponential phase brought about an increase in NAD-GDH activity and, in contrast, a greater and more rapid decrease in NADP-GDH activity (Fig. 18).

These findings suggest a close correlation between the activities of respiration and GDH enzymes as pointed out by De Castro et al. [8,9] in *S. cerevisiae*. Figure 19 shows the activities of the two GDH enzymes in cells growing under anaerobic conditions in Lindegren's medium containing 0.3% glucose. The complex medium was used since cell growth under anaerobic conditions was extremely poor in the synthetic medium even in the presence of added Tween 80 and ergosterol. The activity of NAD-GDH was substantially lower under anaerobic conditions, while there was little difference in NADP-GDH activity between aerobically and anaerobically grown cells. De Castro et al. [8] described that NADP-GDH activity was increased when respiratory competent cells of *S.cerevisiae* were grown anaerobically under derepressed conditions (0.3% glucose or 0.3% galactose). However, the extent of the increments in the activity of the enzyme appeared negligible.

The effect of glucose concentration on cellular activities of GDH enzymes was examined under aerobic conditions in the synthetic medium (Fig. 20). In the absence of added thiamine, NADP-GDH activity was increased and NAD-GDH activity was decreased with increasing concentrations of glucose (0.5 to 10%). The

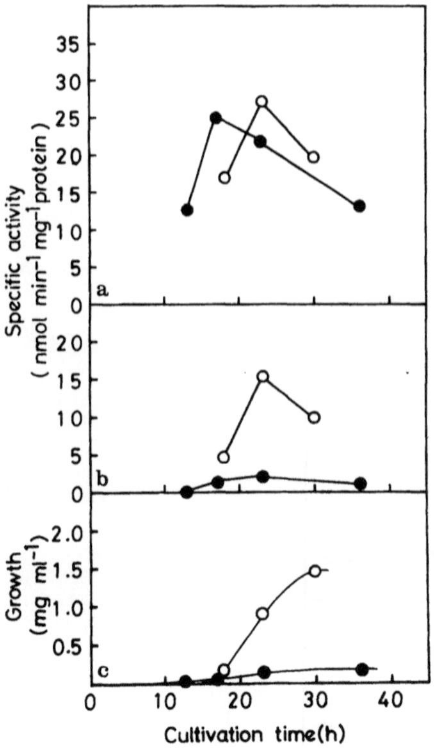

Fig. 19a—c. Effect of anaerobic cultivation on GDH activities [57]. Cells were grown aerobically and anerobically in Lindegren's medium containing 0.3% glucose. GDH enzymes were assayed as described in Fig. 17. (○) Aerobic culture, (●) anaerobic culture. **a** NADP-GDH activity; **b** NAD-GDH activity; **c** growth

increase in NADP-GDH activity brought about by glucose was also observed in the thiamine-supplemented culture. The NADP-GDH activity was markedly higher at 10% glucose than that at 0.5% glucose, especially at the early growth phase when most of the added glucose remained. The thiamine-induced decrease in NAD-GDH activity was not affected by the increase in glucose concentration. Furthermore, at a low concentration of glucose (0.5%), NADP-GDH activity was only slightly increased by thiamine, whereas a marked decrease in respiratory activity by thiamine was observed as at a high concentration range of glucose. These findings indicated that both GDH enzymes are influenced by glucose concentration; NAD-GDH activity is decreased and NADP-GDH activity is increased in the presence of high concentrations of glucose. It was also suggested that the decrease in NAD-GDH activity in the thiamine-grown cells was the result of the thiamine-induced respiratory deficiency and, in contrast, the marked increase in NADP-GDH activity was mainly due to the thiamine-enhanced effect of glucose. These indications were confirmed by the following experiments. When 2% glycerol was used as the carbon source in place of glucose in the synthetic medium, NADP-GDH activity was somewhat decreased by added thiamine (Fig. 21), even though the respiratory activity was lowered. NAD-GDH activity was also slightly lowered, possibly due to the small decrease in respiratory activity. When ALA was added to the synthetic medium containing 5% glucose and thiamine, NAD-GDH activity was markedly increased to a level higher

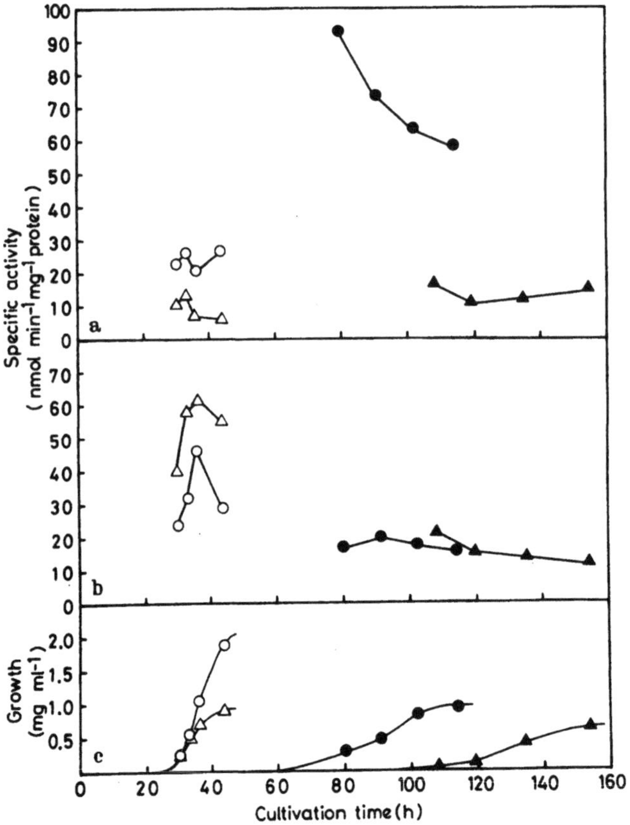

Fig. 20. Effect of glucose on GDH activities [57]. Cells were grown on 0.5 or 10% glucose in the synthetic medium. Conditions were the same as those described in Fig. 17. (○) 10% glucose, without thiamine; (●) 10% glucose, with thiamine; (△) 0.5% glucose, without thiamine; (▲) 0.5% glucose, with thiamine. **a** NADP-GDH activity; **b** NAD-GDH activity; **c** growth

than that of the control cells. The increase in enzyme activity was accompanied by the restoration of respiratory activity [55] as described earlier. However, the thiamine-induced increase in NADP-GDH activity was not influenced by the addition of ALA (Fig. 22).

As described above, in contrast to ALA, pyridoxine which also eliminated the thiamine-induced respiratory deficiency [53] abolished completely the effect of thiamine on NADP-GDH as well as that on NAD-GDH (cf. Fig. 18). There remained a possibility that the thiamine-induced vitamin B_6 deficiency exerted some effect on amino acid metabolism and that GDH enzymes were affected by a resulting alteration in amino acid pools in the cells. We examined intracellular pools of some amino acids and NH_4^+ (Table 16), and the activities of typical vitamin B_6-dependent enzymes, GOT and GPT (Table 17). Thiamine exerted remarkable effects on the pools of methionine, arginine, aspartate, valine, and alanine in the cells harvested in the exponential phase where the effects of thiamine on GDH enzymes were most marked.

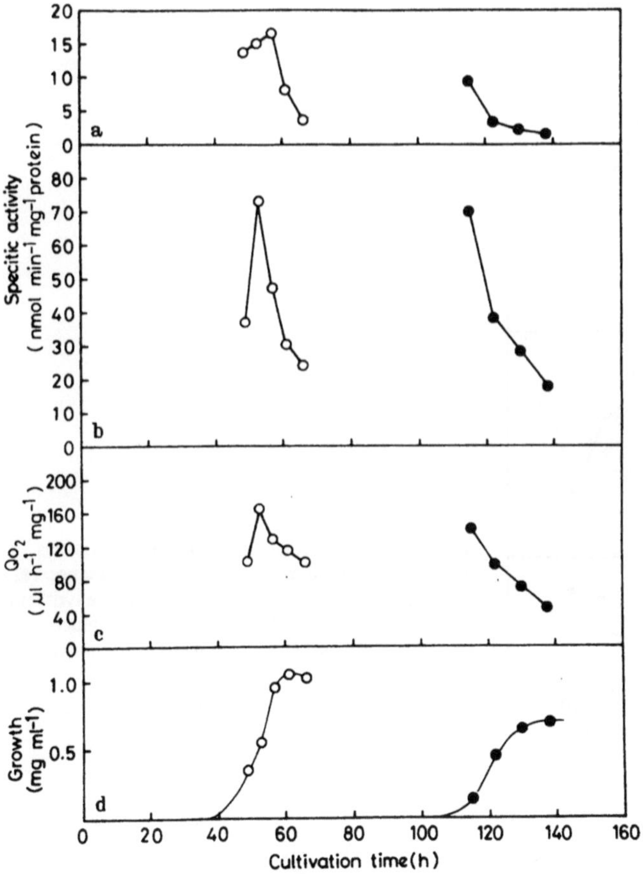

Fig. 21a—d. GDH activities in cells grown aerobically on glycerol [57]. Glycerol (2%) was used in the synthetic medium in place of glucose. (○) Without thiamine, (●) with thiamine. **a** NADP-GDH activity; **b** NAD-GDH activity; **c** respiratory activity (Q_{O_2}); **d** growth

In contrast, the concentrations of glutamate and NH_4^+, which are known to regulate directly NAD-GDH and NADP-GDH activities in yeasts [75, 84], were little affected by thiamine. Holoenzyme activities of both GPT and GOT were reduced by thiamine to one-half those in the control cells. Ratio of holoenzyme activity to total enzyme activity was also decreased by thiamine. This would reflect the thiamine-induced vitamin B_6 deficiency. However, the decreases in the holoenzyme activities were not marked compared with the decrease in ALA synthase activity [55]. Therefore, the effect of pyridoxine, in giving a normal level of NADP-GDH activity, cannot be ascribed to the restoration of amino acid metabolism. It would be very difficult to elucidate the mechanism of the pyridoxine effect on NADP-GDH activity although pyridoxine can be considered to have a close relationship with the glucose effect.

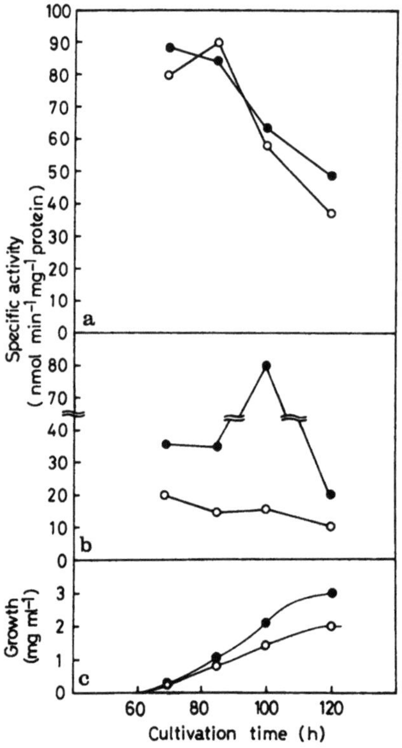

Fig. 22a—c. Effect of ALA on GDH activities [57]. Cells were grown aerobically in the presence (●) or absence (○) of ALA (84 μg · ml^{-1}) in the synthetic medium containing thiamine and 5% glucose. **a** NADP-GDH activity; **b** NAD-GDH activity; **c** growth

Table 16. Effect of thiamine on the content of amino acids and NH_4^+

Amino acid	Content (μmol per 100 mg dry cells)	
	Control cells	Thiamine-grown cells
Aspartate	1.0	3.4
Threonine	4.8	7.0
Serine	3.0	3.0
Glutamate	28.1	36.7
Glycine	1.6	1.2
Alanine	10.3	0.9
Valine	0.8	2.6
Methionine	2.1	12.3
Isoleucine	0.6	0.7
Leucine	0.3	0.5
Lysine	23.3	34.8
Histidine	1.5	2.9
Arginine	5.4	15.0
NH_4^+	3.0	2.7

Cells were harvested at the exponential growth phase where the thiamine effect on GDH enzymes was most marked. Intracellular amino acids and NH_4^+ were extracted and estimated as described in Ref. [57]

The activities of the two GDH enzymes in yeast cells have invariably been reported to change in opposing directions under the environmental conditions employed as if there is some mechanism which regulates these enzymes in a reciprocal manner. We have shown for the first time that the two GDH enzymes can be controlled independently of each other. NAD-GDH is undoubtedly repressed also

Table 17. Effect of thiamine on the activities of glutamate-pyruvate transaminase (GPT) and glutamate-oxalacetate transaminase (GOT)

Cells	Growth of cells (mg ml^{-1})	Specific activity (nmol min^{-1} mg^{-1} of protein)					
		Glutamate-pyruvate transaminase			Glutamate-oxalacetate transaminase		
		Holo-enzyme[a]	Total enzyme[b]	Holo-enzyme/Total enzyme[c]	Holo-enzyme[a]	Total enzyme[b]	Holo-enzyme/Total enzyme[c]
Control	1.19	29.4	167	0.18	131	185	0.71
Thiamine-grown	0.86	15.0	142	0.11	59.6	271	0.21

Cells were harvested at the early growth phase where the difference of NADP-GDH activities between the control cells and the thiamine-grown cells was explicitly observed (cf. Fig. 17). GPT and GOT were assayed as described in Ref. [57].

[a] Assayed without pyridoxal phosphate;

[b] Assayed with pyridoxal phosphate (17 μM);

[c] Ratio of specific activity of the holoenzyme to that of the total enzyme

by glucose, but this may be caused by glucose-induced respiratory repression. It is evident that NADP-GDH activity is independent of respiratory activity under the conditions in this study. De Castro et al. [9] reported that *S. cerevisiae* showed a marked increase in the activity of NADP-GDH when cultivated in the presence of chloramphenicol or other antibiotics that inhibit mitochondrial protein synthesis, although they observed, in their preceeding paper [8], only a very slight increase in anaerobically grown cells as described above. In the synthetic medium used in this study, however, chloramphenicol had no effect on NADP-GDH activity in *S. carlsbergensis* and *S. cerevisica* cells; moreover, a marked increase in NAD-GDH activity was observed in these cells (data not shown). These facts suggest that chloramphenicol exerted some subtle effect on GDH activities in yeast cells, which cannot be explained entirely by respiratory inhibition. It is evident that NADP-GDH is regulated mainly by glucose as judged from our data and other observations [47, 48] which show the rapid degradation and biosynthesis of this enzyme under glucose starvation and refeeding, respectively.

6 Effects of Thiamine and Pyridoxine on Glycolysis and Ethanol Production in *Saccharomyces carlsbergensis* [34]

Regulation between respiration and fermentation (ethanol production) in yeast cells is one of the most important physiological problems. As described in the preceding chapters, the cells of *Saccharomyces carlsbergensis* and other *Saccharomyces* yeasts exhibit very low activity of respiration [51–53, 55] and have altered lipid [50, 58–60] and amino acid [57] metabolisms owing to vitamin B_6 deficiency and/or cytochrome deficiency when grown aerobically with thiamine in a medium free of vitamin B_6. It is generally accepted that the reduction of respiratory activity in yeast cells, as well as in mammalian cells, releases them from the Pasteur Effect and causes an increase in the glycolysis activity and hence, in the case of yeasts increased activity of ethanol production. On the basis of this well-known fact, we investigated the activities of glycolysis and ethanol production in the thiamine-grown cells which are deficient in respiratory activity.

6.1 Glycolysis and Ethanol Production in a Growing Culture

Table 18 shows the effects of thiamine and pyridoxine on ethanol production and energy production in the yeast culture. A small but significant increase in the ethanol yield was observed in the thiamine-supplemented culture. Thiamine caused also a decrease in the molar growth yield. These results indicate that, compared with the control culture, the thiamine-supplemented culture depends on ethanol production for energy production to a large extent. This is consistent with the lowered

Table 18. Ethanol yield and molar growth yield [34]

Culture	Ethanol yield	Molar growth yield
Control	1.53	12.3
Thiamine-grown	1.74	8.6
Thiamine and pyridoxine-grown	1.60	10.1

Aliquots of the culture were withdrawn during cultivation. After the removal of cells, glucose and ethanol concentrations were determined enzymatically by conventional methods. Ethanol yield and molar growth yield were calculated from cell yield and the amount of ethanol accumulated in the culture just at the point at which glucose was completely consumed, and expressed as mmol of ethanol formed per mmol of glucose used and mg of dry cells per mmol of glucose used, respectively

activity of respiration of the thiamine-grown cells. As mentioned earlier, respiratory activity was restored to normal levels upon addition of pyridoxine [52, 53]. In contrast, the effect of thiamine on the ethanol and molar growth yields was not fully abolished by the addition of pyridoxine under the same conditions. It was therefore suggested that the thiamine-induced enhancement of ethanol production cannot be ascribed entirely to the release from the Pasteur Effect.

6.2 Glycolytic and Ethanol Producing Activities of Cells

The effect of thiamine on the cellular activities of glycolysis and ethanol production was examined. Cells grown to the late exponential phase were washed and suspended in 50 mM acetate buffer (pH 5.2) containing glucose at a concentration of 15 mg dry cells per ml. Incubation was carried out at 30 °C with shaking. Aliquots were withdrawn at appropriate time intervals, and the concentrations of glucose and ethanol were assayed.

Table 19. Rates of glucose utilization and ethanol production in resting cells [34)]

Cells	Glucose utilization	Ethanol production
Control	0.109	0.120
Thiamine-grown	0.142	0.227

Cells harvested at the late exponential growth phase were washed with ice-cold water and suspended in 50 mM of potassium acetate buffer (pH 5.2) containing 5% glucose at a final concentration of 15 mg dry cells per ml. The cell suspension was incubated at 30 °C. Aliquots were withdrawn at appropriate time intervals during the incubation. The glucose consumed and the ethanol formed were determined as described in Table 18. The rates of glucose utilization and ethanol production were expressed as μmol of glucose and μmol of ethanol per mg of dry cells, respectively

As shown in Table 19, both the activities were higher in the thiamine-grown cells than in the control cells, indicating that the thiamine-induced respiratory deficiency stimulated glycolysis and ethanol production through the elimination of the Pasteur Effect. On the other hand, the ratio of the increase in the ethanol production rate to the increase in the glucose utilization rate was calculated to be 3.24 (0,107 μmol min^{-1} mg^{-1} dry cells/0,033 μmol min^{-1} mg^{-1} dry cells), and this value was much larger than the theoretical value (=2) which is given by assuming that glucose is converted completely to ethanol. There seems to be a mechanism by which ethanol production is controlled independently of glycolysis. Thiamine would play some important role in this mechanism. A similar experiment was carried out in aerobic and anaerobic cultures. In this case, cells were grown in Lindegren's complex medium due to depressed growth under anaerobic conditions in the synthetic medium as described earlier. As expected, the anaerobically grown cells showed significantly higher activities of glucose utilization and ethanol production than the aerobically grown cells. And, in contrast to the case of the thiamine-grown cells, the ratio of the increase in the ethanol production rate to the increase in the glucose utilization rate was calculated to be 1.98. This value shows a good agreement with the theoretical value (=2), indicating that glucose was converted completely into ethanol under the conditions.

To investigate the mechanism of the thiamine-induced stimulation of glycolysis, several main metabolites accumulated during glucose utilization were assayed after

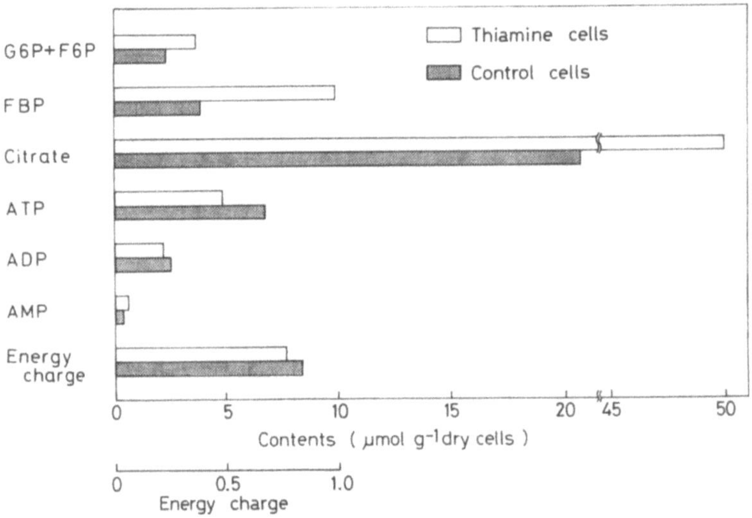

Fig. 23. Effect of thiamine on the accumulation of glycolytic metabolites in cells during incubation with glucose [34]. During the incubation of cells with glucose (cf. Table 19), aliquots were withdrawn and quickly put into ice-cold perchloric acid (1.3 N). After homogenized at 0 °C, the mixture was allowed to stand for 10–20 min at room temperature, followed by centrifugation at 15,000 × g for 10 min. The supernatant was neutralized (pH 6.5) with KOH. After removal of potassium perchloride formed, the metabolite content was determined enzymatically by conventional methods. G6P: glucose-6-phosphate, F6P: fructose-6-phosphate, FBP: fructose-1,6-bisphosphate. Energy charge = (ATP + ADP/2)(AMP + ADP + ATP)

the steady state was reached. The results are depicted in Fig. 23. The thiamine-grown cells (represented as thiamine cells in the figure) accumulated a much higher amount of fructose-1,6-bisphosphate than the control cells, suggesting that the activity of phosphofructokinase, which is believed to be the rate-determining enzyme in glycolysis, is high in the thiamine-grown cells. On the other hand, citrate, which is known to be the most potent inhibitor of phosphofructokinase in yeasts [76], was found in an extremely large amount in the thiamine-supplemented culture, whereas the value of another important factor regulating the enzyme activity, energy charge [79], was not so much influenced by thiamine. Effects of anaerobiosis were also examined in a similar experiment (data not shown). Being consistent with the concept of the Pasteur Effect, anaerobiosis caused a marked decrease in the level of citrate accumulation. Salas et al. [76] described the Pasteur Effect in yeast as being mainly due to the inhibition of phosphofructokinase activity by the accumulated citrate in the culture under aerobic conditions. The inhibition of phosphofructokinase by citrate was confirmed in cell-free extracts from these cells including the thiamine-grown cells. Therefore, the accumulation of citrate in a large amount in the thiamine-grown cells appeared to be in conflict with the enhanced glycolysis. There are numerous effectors of phosphofructokinase other than citrate. The level of phosphate, a positive effector, was also higher in the thiamine-grown cells than in the control cells. However, the enzyme activity was very low when assayed in the presence of all the effectors (positive and negative) at the same concentrations as found in the thiamine-supplemented culture, and this was ascribed to the high level of citrate (data not shown). These results suggested that phosphofructokinase is controlled under the inhibitory effect of citrate in cells growing in the thiamine-supplemented culture. Nevertheless, the cells showed higher activities of glycolysis and ethanol production as compared with the control cells. This conflict can be resolved by assuming that the reaction catalyzed by phosphofructokinase is not the rate-determining step in the yeast, at least under the conditions employed in this study.

Table 20. Levels of glycolytic key enzymes in growing cells [34]

Enzyme	Specific activity (μmol min^{-1} mg^{-1} protein)	
	Control cells	Thiamine-grown cells
Hexokinase	1.2	2.3
Phosphofructokinase	0.39–0.77	0.26–0.83
Pyruvate kinase	12	12

Cells were harvested at the late exponential growth phase. The enzymes were assayed enzymatically by conventional methods in cell-free extracts after removal of low molecular weight substances by gel filtration

As mentioned above, the thiamine-induced stimulation of glycolytic activity would not be explained by the change in the levels of metabolites accumulated in the culture. Then, the levels of the key enzymes in glycolysis were investigated. The enzymes were assayed in cell-free extracts under optimum conditions for each enzyme, so as to compare their levels in the cells. As shown in Table 20, the level of hexokinase in the thiamine-grown cells was twice that in the control cells. It is, however, uncertain that the increase in the level of the enzyme is responsible for the increased activity of glycolysis, since the enzyme activity was markedly lowered upon addition of its effectors to the assay mixture as in the case of phosphofructokinase. The level of phosphofructokinase varied for each experiment and, in some case, the value in the thiamine-grown cells was higher than that in the control cells. However, the activity of the enzyme was affected by its effectors, especially by citrate as mentioned above. The thiamine-grown cells showed the same level of pyruvate kinase as the control cells. The activity of the enzyme in the presence of the effectors was also the same in these different cells. Thus, the mechanism of the thiamine-induced stimulation of glycolysis has not been clarified yet. Very recently, however, a regulatory mechanism of glycolysis at the phosphofructokinase-catalyzed step by fructose-2,6-bisphosphate has been proposed for yeast cells [4,19,45] as well as for mammalian cells [17,18,20,69,74,85–88]. Studies should be done on cellular levels of fructose-2,6-bisphosphate and its effect on the activity of phosphofructokinase in the thiamine-grown cells. Cellular capacity of glucose transport might determine the rate of glycolysis [13,42]. This might be also the case for the thiamine-induced stimulation in glycolysis and ethanol production. If so, glucose transport would be regulated by some metabolism(s) which can be affected by thiamine or thiamine-induced metabolic changes such as respiratory deficiency.

As described above, ethanol production is possibly controlled by thiamine independently of glycolytic activity under the culture conditions employed in this study. To investigate the mechanism of the thiamine-induced stimulation of ethanol production, the levels of pyruvate dehydrogenase complex, pyruvate decarboxylase and alcohol dehydrogenase were determined in both the control and the thiamine-grown cells. As can be seen from Table 21, the thiamine-grown cells exhibited no

Table 21. Levels of enzymes in pyruvate metabolism [34]

Enzyme	Specific activity (μmol min^{-1} mg^{-1} protein)	
	Control cells	Thiamine-grown cells
Pyruvate dehydrogenase complex	0.010	0
Pyruvate decarboxylase	1.7	1.4
Alcohol dehydrogenase	1.0	4.7

Extracts from cells grown to the late exponential phase were used for enzyme assays, which were carried out by conventional methods

appreciable activity of pyruvate dehydrogenase complex which catalyzes the oxidation of pyruvate in mitochondria. On the other hand, the cells had pyruvate decarboxylase at almost the same level as the control cells. The enzyme catalyzes the formation of free acetaldehyde which is reduced to ethanol by the action of alcohol dehydrogenase. The level of the latter enzyme in the thiamine-grown cells was about three times that in the control cells. The markedly lowered activity of pyruvate dehydrogenase and the increased level of alcohol dehydrogenase are consistent with the above-mentioned result in which the increase in ethanol producing activity was much higher than the theoretical value calculated from the increase in glycolytic activity. These facts strongly suggest the presence of an unknown mechanism by which respiration and ethanol production are regulated independently of the cellular level of pyruvate, which has been regarded as the most important factor determining the direction of pyruvate catabolism (oxidation or reduction to ethanol via acetaldehyde) since Holzer [29] described that the affinity for pyruvate of pyruvate decarboxylase is much lower than that of pyruvate dehydrogenase complex.

7 Concluding Remarks

Saccharomyces yeasts have been widely utilyzed for the study of mitochondrial biogenesis although the yeast cells depend mainly on ethanol production for energy production even under aerobic conditions as mentioned in the introductory statement. This is due to the susceptibility of respiratory activity to environmental conditions such as the extent of oxygen supply, the kind and concentration of the carbon source used, the presence of various drugs and so on. By changing such environmental conditions, one can pursue the molecular events occurring in association with the process of mitochondrial development. The thiamine-induced respiratory deficiency and its prevention by pyridoxine and ALA would provide some important information for this field of study. Particularly, the experimental system of respiratory adaptation depending on added pyridoxine or ALA would serve as a unique approach to the study of the mechanism of lipid incorporation into mitochondrial membranes, since our preliminary observations have indicated that the abnormal lipid composition returns to the normal one during the respiratory adaptation. It would be of interest to clarify the process of the restoration of the lipid composition. It is known that addition of unsaturated fatty acids and ergosterol is required for the cultivation of yeasts under anaerobic conditions. It is, therefore, difficult to observe such conversions of lipid composition in the respiratory adaptation system which is usually adopted and which involves the adaptation to oxygen of cells grown anaerobically with added unsaturated fatty acids and ergosterol. Moreover, the effect of hemin, in place of pyridoxine or ALA, for causing the respiratory adaptation and its prevention by protein synthesis inhibitors would be particularly of interest since this suggests the involvement of de novo syntheses of cytochrome proteins, not of heme-synthesyzing enzymes, in the process of the respiratory adaptation.

In addition to the respiratory deficiency and the alteration in lipid composition, thiamine exerted diverse effects on yeast metabolisms and some new findings were presented with respect to the regulation of amino acid metabolism, and of glycolysis and ethanol production. In amino acid metabolism, evidence was provided for the idea that NAD-GDH and NADP-GDH can be regulated independently of each other. The most striking finding is that the ethanol production rate does not depend on the supply of pyruvate, which is determined by glycolytic activity. An entirely new idea was proposed, namely that ethanol-producing activity can control glycolytic activity, at least under the culture conditions employed in this study.

The proposed route (Fig. 24) by which thiamine causes cytochrome deficiency through vitamin B_6 deficiency would be very reasonable, but the mechanisms governing the extraordinary accumulation of thiamine in cells and the occurrence of vitamin B_6 deficiency remain to be elucidated, as well as the in-vivo activity of ALA synthase.

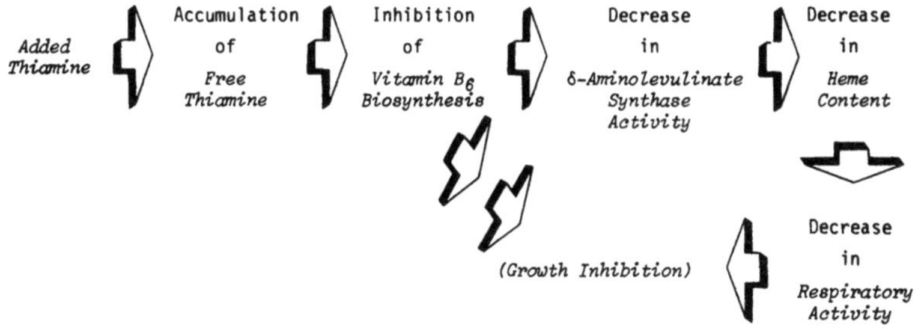

Fig. 24. Proposed mechanism for the regulation of respiration and its related metabolisms in *Saccharomyces* yeasts

The mechanism of the thiamine-induced growth inhibition has not been fully clarified. It would be very difficult to show an exact profile of the metabolic disturbance caused by vitamin B_6 deficiency (Fig. 24). However, systematic screening of active substances in nitrogen compounds for promoting the partially-restored growth by ALA would serve to elucidate this problem.

8 Nomenclature

Acetyl CoA	Acetyl coenzyme A
ADP	Adenosine diphosphate
ALA	δ-Aminolevulinate
AMP	Adenosine monophosphate
ATP	Adenosine triphosphate
C.	*Candida*
CAP	Chloramphenicol
CHI	Cycloheximide
Ethanol yield	mmol of ethanol formed × $mmol^{-1}$ glucose used
F6P	Fructose-6-phosphate
FBP	Fructose-1,6-bisphosphate
G6P	Glucose-6-phosphate
GDH	Glutamate dehydrogenase
K.	*Kluyveromyces*
Molar growth yield	mg of dry cells × $mmol^{-1}$ glucose used
NAD	Nicotinamide adenine dinucleotide
NADH	Nicotinamide adenine dinucleotide (reduced form)
NADP	Nicotinamide adenine dinucleotide phosphate
PLP	Pyridoxal phosphate
Q_{O_2}	$\mu l\ h^{-1}\ mg^{-1}$ dry cells
S.	*Saccharomyces*
Sch.	*Schizosaccharomyces*
Succinyl CoA	Succinyl coenzyme A
TPP	Thiamine pyrophosphate

9 References

1. Aoki, Y. et al.: J. Clin. Invest. *53*, 1326 (1974)
2. Aoki, Y.: J. Biol. Chem. *253*, 2026 (1978)
3. Atkin, L. et al.: Ind. Eng. Chem., Anal. Ed. *15*, 141 (1943)
4. Avigad, G.: Biochem. Biophys. Res. Commun. *102*, 985 (1981)
5. Cartledge, T. G. et al.: Biochem. J. *130*, 739 (1972)
6. Chiao, J. S. et al.: Arch. Biochem. Biophys. *64*, 115 (1956)
7. Clark-Malker, G. D. et al.: J. Cell Biol. *34*, 1 (1967)
8. De Castro, I. N. et al.: Eur. J. Biochem. *16*, 567 (1970)
9. De Castro, I. N. et al.: Can. J. Biochem. *3*, 109 (1974)
10. De Deken, R. H.: J. Gen. Microbiol. *44*, 149 (1966)
11. Doelle, H. W. et al.: Adv. Biochem. Eng. *23*, 1 (1982)
12. Ephurussi, B. et al.: Biochim. Biophys. Acta *6*, 256 (1950)
13. Fiechter, A. et al.: Adv. Microb. Physiol. *22*, 123 (1981)

14. Fryberg, M. et al.: J. Am. Chem. Soc. *95*, 5747 (1973)
15. Fryberg, M. et al.: Arch. Biochem. Biophys. *160*, 83 (1974)
16. Fryberg, M. et al.: ibid. *173*, 171 (1975)
17. Furuya, E. et al.: Proc. Natl. Acad. Sci. U.S.A. *77*, 5861 (1980)
18. Furuya, E. et al.: J. Biol. Chem. *256*, 7109 (1981)
19. Furuya, E. et al.: Biochem. Biophys. Res. Commun. *105*, 1519 (1980)
20. Furuya, E. et al.: Proc. Natl. Acad. Sci. U.S.A. *79*, 325 (1982)
21. Gibson, K. D. et al.: Biochem. J. *70*, 71 (1958)
22. Gibson, K. D.: Biochim. Biophys. Acta *28*, 451 (1958)
23. Gollub, E. G. et al.: J. Biol. Chem. *252*, 2846 (1977)
24. Gross, M. et al.: Proc. Natl. Acad. Sci. U.S.A. *69*, 1565 (1972)
25. Hamilton-Miller, J. M. T.: Adv. Appl. Microbiol. *17*, 109 (1974)
26. Hata, S. et al.: Biochem. Biophys. Res. Commun. *103*, 272 (1981)
27. Hayashi, N. et al.: Arch. Biochem. Biophys. *131*, 83 (1969)
28. Henderson, G. B. et al.: J. Bacteriol. *133*, 1190 (1978)
29. Holzer, H.: Cold Spring Harbor Symp. Quant. Biol. *26*, 277 (1961)
30. Huang, M. et al.: Biochim. Biophys. Acta *114*, 434 (1966)
31. Irving, E. A. et al.: J. Biol. Chem. *244*, 60 (1969)
32. Iwashima, A. et al.: Biochim. Biophys. Acta *330*, 222 (1973)
33. Kamihara, T. et al.: Hakko to Kogyo (Fermentation and Industry) *35*, 374 (1977)
34. Kamihara, T. et al.: Adv. Biotechnol. *2*, 225 (1981)
35. Kakiuchi, Y.: Bitamin (Vitamins) *40*, 454 (1969)
36. Kawasaki, C. et al.: ibid. *40*, 379 (1969)
37. Kawasaki, T. et al.: Arch. Biochem. Biophys. *131*, 223 (1969)
38. Kawasaki, T. et al.: ibid. *142*, 163 (1971)
39. Kikuchi, G. et al.: J. Biol. Chem. *233*, 1214 (1958)
40. Kitsutani, S. et al.: Plant Cell Physiol. *11*, 551 (1970)
41. Krebs, H. A.: Essays in Biochemistry *8*, 1 (1972)
42. Lagunas, R. et al.: J. Bacteriol. *152*, 19 (1982)
43. Lascelles, J.: Tetrapyrrole biosynthesis and its regulation, p. 45, New York: Benjamin 1964
44. Laver, W. G. et al.: Biochem. J. *70*, 4 (1958)
45. Lederer, B. et al.: Biochem. Biophys. Res. Commun. *103*, 1281 (1981)
46. Linnane, A. W. et al.: The effect of chloramphenicol on the differentiation of the mitochondrial organelle, in: Aspects of yeast metabolism Mills, A. K. (ed.), p. 217, Oxford-Edinburgh: Blackwell 1967
47. Mazon, M. J.: J. Bacteriol. *133*, 780 (1978)
48. Mazon, M. J. et al.: ibid. *139*, 686 (1979)
49. Marzluf, G. A.: Microbiol. Rev. *45*, 437 (1981)
50. Nagai, J. et al.: Biochem. Biophys. Res. Commun. *60*, 555 (1974)
51. Nakamura, I. et al.: ibid. *59*, 771 (1974)
52. Nakamura, I. et al.: FEBS Lett. *62*, 354 (1976)
53. Nakamura, I. et al.: Arch. Microbiol. *127*, 47 (1980)
54. Nakamura, I. et al.: Seikagaku (Biochemistry) *53*, 795 (1981)
55. Nakamura, I. et al.: J. Bacteriol. *147*, 954 (1981)
56. Nakamura, I. et al.: J. Gen. Microbiol. *128*, 2601 (1982)
57. Nakamura, I. et al.: ibid, *129*, 945 (1983)
58. Nishikawa, Y. et al.: Biochem. Biophys. Res. Commun. *59*, 777 (1974)
59. Nishikawa, Y. et al.: Biochim. Biophys. Acta *486*, 483 (1977)
60. Nishikawa, Y. et al.: ibid. *531*, 86 (1978)
61. Nishimune, T.: Bitamin (Vitamins) *47*, 221 (1973)
62. Ohba, M. et al.: Biochem. Biophys. Res. Commun. *85*, 21 (1978)
63. Okada, H. et al.: Biochim. Biophys. Acta *82*, 538 (1964)
64. Oshiba, K. et al.: Bitamin (Vitamins) *33*, 47 (1966)
65. Oshino, N. et al.: Biochim. Biophys. Acta *128*, 13 (1966)
66. Pachecka, J. et al.: Bull. Acad. Pol. Sci., Serie des sciences biologiques C. L. II. *19*, 17 (1971)
67. Patterson, G. W.: Anal. Chem. *43*, 1165 (1971)
68. Pflug, W. et al.: Hoppe-Seyler's Z. Physiol. Chem. *359*, 559 (1978)

69. Pilkis, S. J. et al.: J. Biol. Chem. *256*, 3171 (1981)
70. Plant, G. W. E. et al.: Ann. Rev. Biochem. *43*, 899 (1974)
71. Rabinowitz, J. C. et al.: Arch. Biochem. Biophys. *33*, 472 (1951)
72. Ramaiah, A.: Curr. Topics Cell. Reg. *8*, 297 (1974)
73. Rattray, J. B. M. et al.: Bacteriol. Rev. *39*, 197 (1975)
74. Richards, C. S. et al.: Biochem. Biophys. Res. Commun. *100*, 1673 (1981)
75. Roon, R. J. et al.: J. Bacteriol. *116*, 367 (1973)
76. Salas, M. L. et al.: Biochem. Biophys. Res. Commun. *19*, 371 (1965)
77. Schroer, R. A. et al.: Proc. Soc. Exp. Biol. Med. *142*, 369 (1973)
78. Schulman, M. P. et al.: J. Biol. Chem. *226*, 181 (1957)
79. Shen, L. C. et al.: Biochemistry *7*, 4041 (1968)
80. Simpson, D. M. et al.: J. Biol. Chem. *255*, 1630 (1980)
81. Sugiura, T. et al.: Biochim. Biophys. Acta *115*, 267 (1966)
82. Suzuoki, J.: J. Biochem. *42*, 27 (1955)
83. Tamura, Y. et al.: Arch. Biochem. Biophys. *175*, 284 (1973)
84. Thomulka, K. W. et al.: J. Bacteriol. *109*, 25 (1972)
85. Uyeda, K. et al.: J. Biol. Chem. *256*, 8679 (1981)
86. Van Schaftingen, E. et al.: Biochem. Biophys. Res. Commun. *96*, 1524 (1980)
87. Van Schaftingen, E. et al.: Biochem. J. *192*, 881 (1980)
88. Van Schaftingen, E. et al.: Biochem. Biophys. Res. Commun. *103*, 362 (1981)
89. Vell, G. J. et al.: J. Biol. Chem. *255*, 3042 (1982)
90. Wintrobe, M. M.: Harvey Lectures *45*, 87 (1950)
91. Yamauchi, K. et al.: J. Biol. Chem. *255*, 1746 (1980)

Anaerobic Wastewater Treatment

H. Sahm
Institut für Biotechnologie der Kernforschungsanlage Jülich, D-5170 Jülich, FRG

The article reviews the present understanding of bacterial populations involved in anaerobic degradation of organic material into methane and CO_2 (biogas); furthermore some recent process developments for anaerobic wastewater treatment are described. It could be demonstrated that at least three groups of bacteria are involved in methanogenesis. Hydrolytic and acidogenic bacteria first decompose the organic material into various organic acids, alcohols, hydrogen and CO_2. The second group of bacteria convert these metabolites into acetic acid, hydrogen and CO_2, which are then utilized by the methanogenic bacteria to produce biogas.

On an industrial scale, this process has been used for more than 50 years in the stabilization of sewage sludge from municipal water treatment plants. Since recent developments have markedly reduced the retention time for anaerobic fermentation*, this process has gained increasing interest in the treatment of high strength wastewater. The anaerobic digestion has two unique advantages over the aerobic biological treatment systems: no energy is necessary for aeration and the organic pollutants are mainly converted into biogas which can be used as a fuel. Until now this anaerobic process is successfully used for treatment of wastewaters from sugar factories, potato-processing industry and breweries. However, fundamental studies have shown that anaerobic treatment offers an enormous potential for the removal of organic materials from a lot of different wastewaters. Therefore, in future the application of anaerobic digestion for wastewater treatment will increase.

---

* In this article the word fermentation is used in the strict sence of Pasteur as anaerobic catabolism.

1 Introduction

Methane is produced everywhere in nature where organic material is degraded by microorganisms in the absence of oxygen, sulfate and nitrate; under these conditions CO_2 is the only available electron acceptor [1]. The discovery by the Italian physicist A. Volta in 1776 that "combustible air" was being formed in the sediments of rivers or lakes rich in settled organic compounds led to subsequent discoveries by Béchamp, Popoff, Tappeneiner, Hoppe Seyler, Söhngen, and Omelianski, that defined a microbial basis for the origin of methane gas [2].

In recent years atmospheric chemists measured about 500–800 million tons of biologically generated methane released into the atmosphere per year [3]. Biogenic methane is the major source of atmospheric methane and is quantitatively similar to the output from natural gas wells [4]. This corresponds to as much as 0.5% of the total annual production of dry organic matter synthesized by photosynthesis. However, environmental studies indicate that the true biological activity of methanogenic bacteria as estimated above is masked by methane utilizing bacteria, which oxidize methane to CO_2. Hence much of the methane produced at the bottom of a lake does not reach the atmosphere, but is oxidized microbiologically to CO_2. Thus this anaerobic process forms an essential link in the natural carbon cycle. Partial methane formation occurs also in the gastro-intestinal tract of ruminants; one cow produces 200 or more liters of methane per day which are removed by belching. This organic carbon excretion constitutes an energy loss of approximately 8–10%. Methane is therefore one of the most abundant organic compounds on this planet, and this is reflected by the abundance of methanogens and methanotrophs in the environment.

On an industrial scale, microbial methane formation has been used since the beginning of this century for the stabilization of sewage sludge from wastewater plants [5]. Anaerobic digestion is well suited to reduce the organic matter present in the sludge and to release the bound water so that the sludge can be concentrated to a form suitable for incineration or mechanical disposal. The methane gas from these anaerobic digestion can be used for the operation of the wastewater treatment plants, and in well designed plants a respectable amount of surplus energy remains for other consumers. In recent years anaerobic treatment has also become accepted as an effective means of treating high strength waste-waters with the following advantages over the the aerobic biological treatment systems: 1. no energy is necessary for aeration, 2. the organic pollutants are converted almost quantitatively to a high

Table 1. Comparison of the carbon- and energy-balances between aerobic and anaerobic microbial degradation processes

	Aerobic conditions	Anaerobic conditions
Carbon balance	About 50% is converted into biomass and 50% into CO_2	About 95% is decomposed into biogas and 5% is incorporated into biomass
Energy balance	Approx. 60% is stored in the large amount of new cells formed and 40% is lost as process heat	Almost 90% of the energy in the organic material can be recovered in the biogas, 5–7% is used for growth of cells and 3–5% is wasted as heat

energy fuel (biogas), and 3. only negligible excess of microbial biomass (sludge) is formed (Table 1).

In this article the microbiology of anaerobic degradation and methanogenesis and some recent developments for anaerobic wastewater treatment will be described and discussed.

2 Microbiology of Anaerobic Digestion

Present understanding of bacterial populations involved in anaerobic digestion is rather limited and is based on analysis of bacteria isolated from sewage sludge digesters or from the rumen of some animals [5,6]. Until recently this methane fermentation was discussed in terms of two groups of bacteria or stages of degradation, the acid-forming and methane-forming stages [2]. The acid-forming stage involved the acidogenic bacteria which hydrolyze polymers and convert the products to organic acids, alcohols, CO_2 and H_2. The methane-forming stage involved the methanogenic bacteria which catabolized these compounds to the final products CH_4 and CO_2. However, methanogenic bacteria are not able to catabolize alcohols other than methanol or organic acids other than acetate and formate [7,8], that is why at least three groups of bacteria decompose organic material into methane and CO_2 (Fig. 1). The first group, as before, involves the hydrolytic and acidogenic bacteria. The second metabolic group, collectively called the H_2-producing acetogenic bacteria converts the various products formed by the first group into H_2, CO_2 and acetate. The third group comprises the methanogenic bacteria which utilize H_2, CO_2, and acetate in production of the final products CH_4 and CO_2. In the digestive tract of living organisms the methane formation involves only the first group — the hydrolytic and acidogenic bacteria — and the H_2-utilizing methanogenic bacteria [6]. The organic acids etc. are adsorbed from the tract and utilized as major carbon and energy sources in animals. This is probably accomplished by the relatively short hydraulic residence time in the gastrointestinal tract (less than one day). The acetogenic bacteria which catabolize fatty acids and the acetophilic methanogens do not grow fast enough to be maintained in the system.

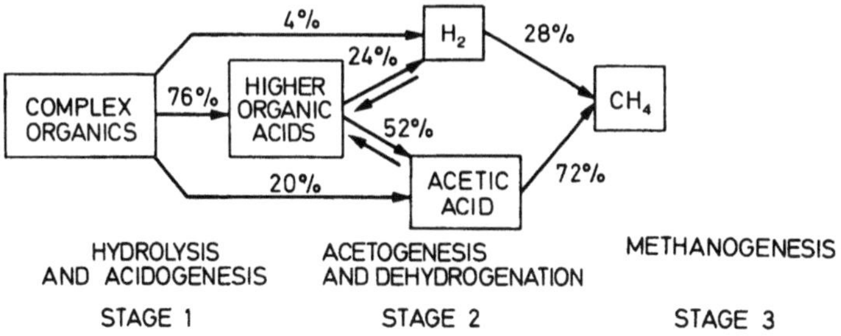

Fig. 1. The three stages of the methane fermentation. Percentages represent the flow of electrons from organic compounds to methane [40]

2.1 Hydrolytic and Acidogenic Bacteria

A complex consortium of microorganisms participates in the hydrolysis and fermentation of organic material. Populations of 10^8 to 10^9 hydrolytic bacteria per ml of mesophilic sewage sludge or 10^{10}–10^{11} hydrolytic bacteria per gram of volatile solids have been found [9]. Most of the bacteria among the predominating flora are strict anaerobes such as *Bacteroides*, *Clostridia*, *Bifidobacteria* and other gram-positive and gram-negative rods. Furthermore, some facultative anaerobes such as *Streptococci* and *Enterobacteriaceae* were detected. First of all the polymeric organic material such as polysaccharides, fats and proteins are hydrolyzed by extracellular enzymes, excreted by several bacteria. Although most of the biopolymers are easily degradable, the cellulose of highly lignified plant material (straw, wood, etc.) has been shown to be very resistant to hydrolysis [10]. The sugars and amino acids formed are then taken up by the bacteria and fermented mainly to acetate, propionate, butyrate, H_2, CO_2, lactate, valerate, ethanol, ammonia and sulfide (Fig. 2) [11,12]. Succinate produced by many bacteria is decarboxylated by others to yield propionate and CO_2 [13].

The concentration of H_2 plays a central role in controlling the proportions of the various products from acidogenic bacteria [6,11,14]. As shown in Fig. 3 the free energy change for the oxidation of $NADH_2$ to NAD and H_2 is negative at a partial pressure of H_2 below 10^{-3} atm. Thus in mixed cultures, where p_{H_2} is kept low by the action of H_2-utilizing methanogenic bacteria, the fermentative metabolism is shifted to the production of more hydrogen and less reduced organic compounds. In an efficiently

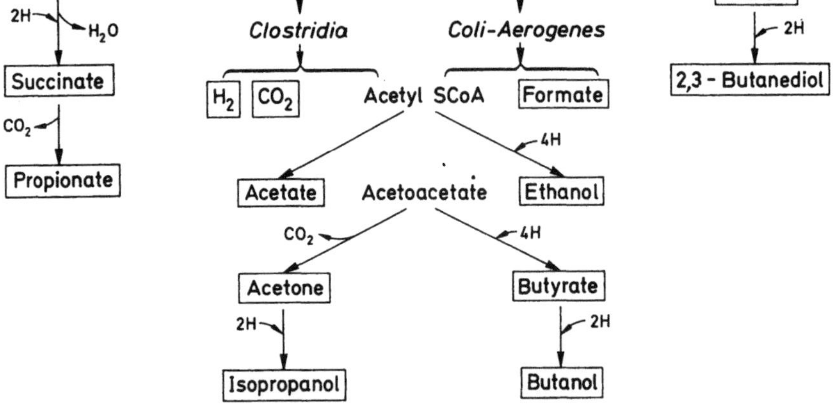

Fig. 2. Bacterial fermentation products of pyruvate. Pyruvate is formed by the catabolism of carbohydrates

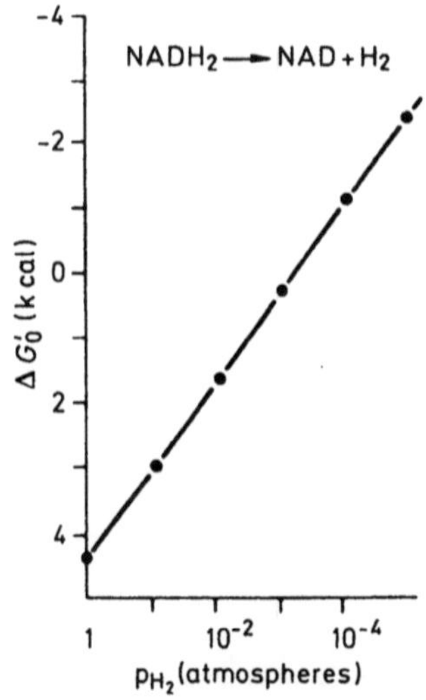

Fig. 3. The free energy change for the oxidation of $NADH_2$ to NAD and H_2 depend on the partial pressure of hydrogen [12]

operating methanogenic ecosystem, i.e. when the partial pressure of H_2 is maintained at a very low level, most of the carbohydrate is converted into acetate, CO_2 and H_2 and without major production of other fatty acids. However, when H_2 concentration increases as, for example, when stress is put on the methanogenic system by shortening the retention time or overloading the system with degradable organic matter, there is an increasing tendency for the $NADH_2$ generated in fermentation to be utilized in the formation of reduced products especially propionic, butyric, valeric, caproic, and lactic acids [11,14].

2.2 H_2-Producing Acetogenic Bacteria

Until 1967 is was thought that the methanogenic bacteria use directly the products of the acidogenic bacteria. However, Bryant et al. [7] showed that *Methanobacillus omelianskii* was not a pure but a mixed culture containing synergistic association of two species. Their production of methane from ethanol follows Eq. (1):

$$2\,CH_3CH_2OH + CO_2 \rightarrow 2\,CH_3COOH + CH_4 \qquad (1)$$
$$\Delta G_0' = -132{,}6\ \text{kJ}$$

One of the species, the acetogenic S organism, oxidizes ethanol to acetate and hydrogen according to Eq. (2):

$$CH_3CH_2OH + H_2O \rightarrow CH_3COOH + 2\,H_2\ ; \qquad (2)$$
$$\Delta G_0' = +6.2\ \text{KJ}\ .$$

Since $\Delta G_0'$ at pH 7.0 for the conversion of ethanol to acetate and H_2 is positive, the equilibrium of the reaction is to the left and the organism is not able to grow on ethanol. However, as the partial pressure of H_2 is lowered, the $\Delta G_0'$ becomes

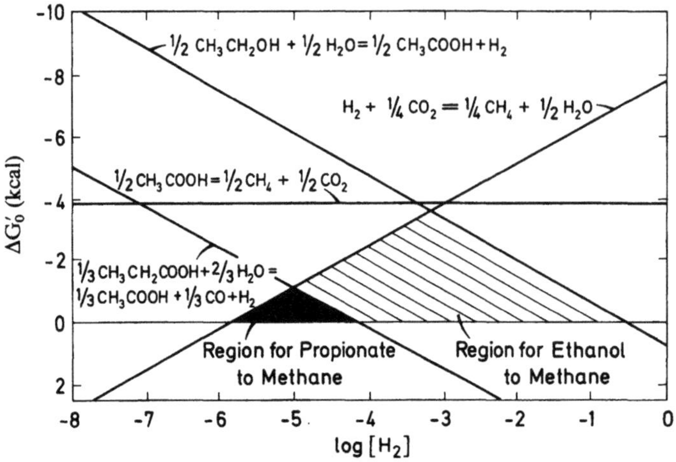

Fig. 4. Effect of hydrogen partial pressure on the free energy of conversion of ethanol, propionate, acetate and hydrogen during methane fermentation [40]

progressively more negative (Fig. 4) and the organism will then be able to grow on ethanol. The second bacterium in this mixed culture was a methanogen which could not use ethanol but used H_2 according to Eq. (3):

$$4\,H_2 + CO_2 \rightarrow CH_4 + 2\,H_2O\,; \qquad (3)$$
$$\Delta G_0' = -138{,}9\ \text{KJ}\,.$$

Propionate, longer carbon-chained fatty acids, alcohols, some aromatic compounds such as benzoate and other organic acids produced by the acidogenic bacteria are also degraded by acetogenic bacteria in association with methanogenic bacteria [15,17]. The $\Delta G_0'$ for the degradation of propionate according to Eq. (4):

$$CH_3CH_2COOH + 2\,H_2O \rightarrow CH_3COOH + CO_2 + 3\,H_2 \qquad (4)$$
$$\Delta G_0' = +48.1\ \text{KJ}$$

is much more positive than for ethanol oxidation. Therefore it was much more difficult to isolate the bacterium effecting this reaction. The maintenance of an extremely low partial pressure of H_2 in the environment by the methanogenic bacteria is essential for the acetogenic and H_2-producing bacteria. Fig. 4 illustrates the relationship between the hydrogen partial pressure and the free energy available to the hydrogen producing bacteria and hydrogen-consuming methanogens. In order for energy to be available to the organism oxidizing propionic acid to acetic acid and hydrogen, the partial pressure of H_2 cannot exceed about 10^{-6} bars [11]. This low H_2-pressure can be obtained by an intensive cell contact between the acetogenic and methanogenic bacteria (interspecies-hydrogen-transfer). Under these conditions, the energy available to the hydrogen-oxidizing bacteria is reduced considerably from what it would be at partial pressure near one atmosphere. This results in much lower bacterial yields per mole of hydrogen gas oxidized, as confirmed by the low overall growth yields measured by Lawrence et al. for complete methane fermentation of propionate and other fatty acids [18].

Populations of 4.2×10^6 hydrogen producing acetogenic bacteria per ml of sewage sludge have been reported [19]. These bacteria have not been generically identified or physiologically well characterized. A new genus and species of a nonmotile gram-negative rod, *Syntrophobacter wolinii*, degrades propionate to acetate, CO_2 and H_2 only in coculture with an H_2-utilizing organism [20]. Furthermore, an anaerobic, nonphototrophic bacterium, *Synthrophomonas wolfii*, that β-oxidizes saturated fatty acids in a syntrophic association with H_2-using bacteria was isolated [21]. In the absence of sulfate members of the genus *Desulfovibrio* it can oxidize ethanol and lactate when combined with methanogenic bacteria [8, 22].

2.3 Methanogenic Bacteria

These bacteria are strict anaerobes; they require a lower redox potential (E_h) for growth than most other anaerobic bacteria. Since the limiting E_h for growth is -330 mV, which corresponds to a concentration of 1 molecule of oxygen in about 10^{56} l water [23], oxygen is a potent inhibitor of methanogenesis; nevertheless, the methanogens may not be killed by exposure to high concentrations of oxygen provided they are subsequently cultured under conditions of low redox potential [24]. This perhaps explains their widespread distribution in nature. Thus the methanogens depend on the fermentative bacteria to produce anaerobiosis and to maintain a low redox potential.

Special anaerobic techniques were developed to isolate these methanogenic bacteria in pure culture and to grow them under defined conditions [23,25]. The methanogens that have been isolated and studied so far in pure cultures are shown in Table 2 [26]. These bacteria can be gram positive or gram negative and they have quite different cell shapes. For example they include large sarcinae, coccus groups similar to micrococci and streptococci, long cylindrical rods, long curved rods with tufts of flagella and short rods with a single polar flagellum.

The methanogenic bacteria are physiologically united by their requirement to form methane as a final product of energy metabolism. As shown in Table 2 these bacteria have a very limited substrate spectrum. Most species oxidize H_2 and reduce CO_2 to form methane as their preferred pathway of methanogenesis. Since the change in free energy of this reaction is very negative [11], the bacteria obtain their energy from this process. The methanogens have a great affinity for H_2; for example Hungate et al. found a K_m value for utilization of H_2 for CH_4 production in the rumen to be 10^{-6} M [27]. Many of the methanogenic species are also able to utilize formate which is catabolized in a manner similar to H_2:

$$4\,HCOOH \rightarrow CH_4 + 3\,CO_2 + 2\,H_2O$$
$$\Delta G'_0 = -111\,KJ\,. \tag{5}$$

Althouth about 70% of the methane produced from organic matter in nature is produced via the methyl group of acetate (Fig. 1), only a few species are able to

Table 2. Characteristics of methanogenix species in pure Culture [26]

Species	Morphology	Substrates	Cell wall composition
Methanobacterium	Long rods	H_2, formate	Pseudomurein
formicium	to	H_2	
bryantii	filaments	H_2	
thermoautotrophicum			
Methanobrevibacter	Lancet-shaped	H_2, formate	Pseudomurein
ruminantium	cocci	H_2, formate	
smithii	short rods	H_2	
arboriphilus			
Methanococcus	Motile irregular	H_2, formate	Polypeptide
vannielii	small	H_2, formate	subunits
voltae	cocci	H_2, formate	
thermolithotrophicus	Pseudosarcina	H_2, methanol,	
mazei		methylamines,	
		acetate	
Methanomicrobium	Motile short	H_2, formate	Polypeptide
mobile	rods		subunits
Methanobacterium	Motile irregular	H_2, formate	Polypeptide
cariaci	small cocci	H_2, formate	subunits
marisnigri			
Methanospirillum	Motile regular	H_2, formate	Polypeptide
hungatei	curved rods		
Methanosarcina	Irregular cocci	H_2, acetate	Heteropoly-
barkeri	as single cells	methanol	saccharide
	packets, pseudo-	methylamines	
	parenchyma		
Methanothrix	Rods to	Acetate	No muramic
soehngenii	long filaments		acid
Methanothermus	Non-motile	H_2	Pseudomurein
fervidus	rods		

degrade acetate into CH_4 and CO_2. Thus far, only three aceticlastic methanogenic species have been isolated in pure culture. They are *Methanosarcina barkeri*, *Methanococcus mazei*, and *Methanothrix soehngenii*. Since the free-energy change of the conversion of acetate into methane and CO_2 is very small:

$$CH_3COOH \rightarrow CH_4 + CO_2 \qquad \Delta G_0' = -32\,KJ\,, \tag{6}$$

these methanogens grow very slowly on this substrate. *M. soehngenii* has a generation time of 10 d or more, but since it exhibits a relatively low K_s-value for acetate (0.7 mMol l^{-1}), it can compete with other strains at lower acetate concentrations [28]. Most of the physiological studies have been done on *M. barkeri* strains which grow much faster on acetate (generation time 2–3 d) and have a K_s-value for acetate of 5 mMol l^{-1}. It was found that hydrogen exerts a regulatory effect on the aceticlastic reaction and prevents all strains of *Methanosarcina* from metabolizing acetate[29].

Methanosarcina barkeri and *Methanococcus mazei* are among the highly specialized methanogenic bacteria, the most versatile with respect to the number of substrates utilized. In addition to acetate and H_2 they also use methanol and methylamines [30,31]:

$$^4/_3\ CH_3OH \rightarrow CH_4 + {}^1/_3\ CO_2 + {}^2/_3\ H_2O$$
$$\Delta G_0' = -106\ KJ \quad (7)$$

$$^4/_3\ CH_3NH_2 + {}^2/_3\ H_2O \rightarrow CH_4 + {}^1/_3\ CO_2$$
$$\Delta G_0' = -77\ KJ\ . \quad (8)$$

The older literature indicates that also many other compounds such as propionate, butyrate, valerate and capronate can be metabolized by methanogens [2]. However, since excellent techniques have not been successful in isolating such strains in pure culture, it seems that these results were obtained with mixed cultures containing acetogenic and methanogenic bacteria together [14].

These methanogenic bacteria differ from the classical bacteria and the eucaryotic organisms with respect to several biochemical characteristics, as:

1. Muramic acid which is characteristic for the wall of typical bacteria is not present in the cell walls of the methanogens [32]. As shown in Table 2 the *Methanobacteriaceae* and *Methanobrevibacteriaceae* have a cell wall polymer called pseudomurein which resembles the murein, but it contains L-talosaminuronic acid instead of muramic acid and the amino acids are exclusively in the L-configuration. All the other methanogens possess envelopes composed of glycoprotein or protein subunits. Therefore these

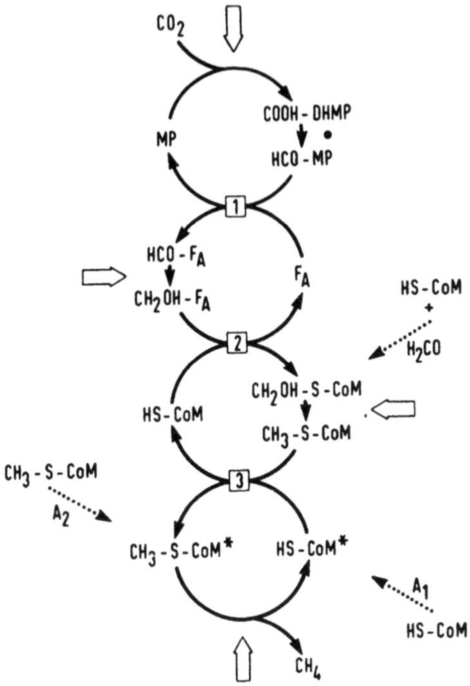

Fig. 5. Tentative scheme for the reduction of CO_2 to methane. The C_1-carriers are methanopterin (MP), 7,8-dihydromethanopterin (DHMP), a coenzyme F_A, and coenzyme M [34]

bacteria are resistant to the cell wall active antibiotics such as: penicillin, D-cycloserine, vancomycin and cephalosporin.

2. The composition of the lipids differs markedly from that of the typical bacteria. The main components of the non-polar lipid fraction are acyclic, isoprenoid hydrocarbons, mainly phytanyl (C_{20}), pentaisoprenyl (C_{25}) and squalenyl (C_{30}) compounds. The polar lipid fraction consists of glycerol ethers of the saturated C_{20} and C_{40} isoprenoid hydrocarbons, the two main components are: biphytanylglycerol diether and dibiphytanylglycerol tetraether [33].

3. Several new coenzymes and factors have been discovered in methanogens [34]: Coenzyme M ($HS—CH_2—CH_2—SO_3H$), is involved in the last step of methanogenesis. Coenzyme F_{420}, a 5-deazaisoalloxacine derivative, acts as an electron carrier. There is some evidence that methanogenic bacteria can be quantitatively assessed in a mixed culture by the spectrofluorimetric determination of this blue-green fluorescent compound [35]. Factor F_{430}, the prosthetic group of methyl CoM reductase, is a nickel tetrapyrrole; therefore the growth of methanogens depends on nickel [36].

4. Application of sequencing techniques for measuring the phylogenetic relationship between different organisms demonstrated, that the ribosomal RNA sequences of the methanogens are quite different from those that of the classical bacteria (eubacteria) [37]. Therefore the methanogenic bacteria belong to the *Archaebacteria*, which is an ancient group phylogenetically distinct from the typical procaryotes. The *Archaebacteria* include, in addition to the methanogens, the extreme halophile (*Halobacteriaceae*) and the thermoacidophiles, *Sulfolobus* and *Thermoplasma*[25].

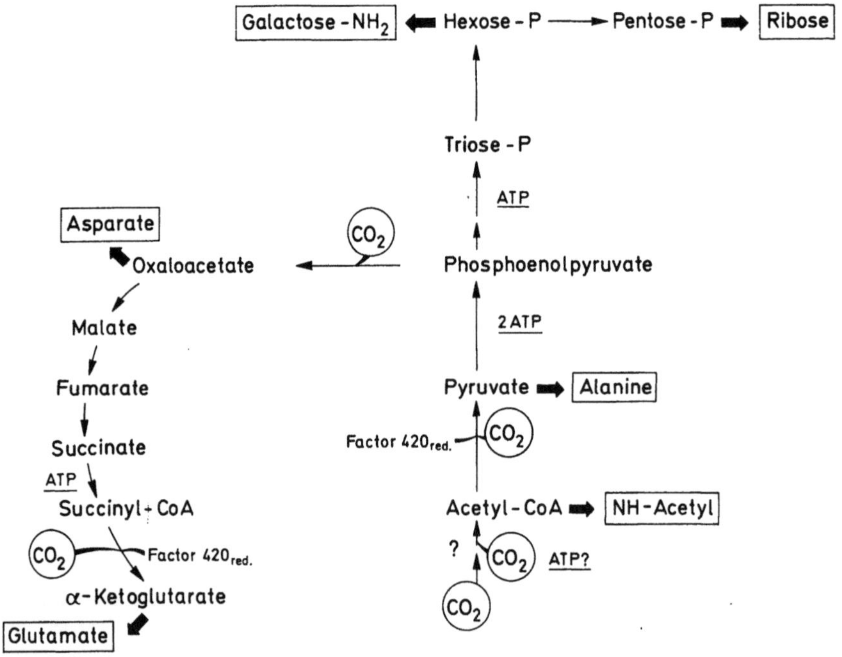

Fig. 6. Proposed autotrophic CO_2 assimilation pathway in *Methanobacterium thermoautotrophicum* [39]

Carbon dioxide is reduced to methane through stages equivalent to formate, formaldehyde and methanol but these compounds are not free intermediates [2]. As shown in Fig. 5, a tentative scheme of methanogenesis, the C_1-units are fixed to different carriers during the successive reduction [34]. CO_2 is bound to methanopterin, which contains a pterin group, glutamic acid, an enamine moiety, cyclohexanetriol, talosamine and dihydroxybutyric acid. By the reduction of CO_2 carboxy-7,8-dihydromethanopterin is formed, which can be converted to formylmethanopterin. The accepting coenzyme of the C_1-unit at the formyl level is still unknown. In the last two steps coenzyme M seems to be involved since CH_2OH—S—CoM and CH_3—S—CoM can be converted to methane. Thus a unique set of coenzymes and enzymes is involved in the methanogenesis.

Most methanogenic bacteria grow on CO_2 and H_2 as sole carbon and energy sources. Labelling data and enzymatic studies indicate that autotrophic CO_2 fixation in these bacteria does not proceed via the Calvin cycle [38]. It was found that acetyl-CoA is the key intermediate of CO_2 assimilation in *Methanobacterium thermoautotrophicum* (Fig. 6) [39]. Acetyl-CoA seems to be synthesized from 2 CO_2 by an as yet unknown mechanism. CO_2 is fixed via the carboxylation of acetyl-CoA to pyruvate, of phosphoenolpyruvate to oxalacetate and of succinyl CoA to α-ketogluterate.

3 Anaerobic Treatment Systems

3.1 Conventional Digestion Tanks

These are digesters normally used for sewage sludges, animal slurries and other very concentrated wastewater (COD higher than 30000 mg l^{-1}).

One century ago (1881), an air-tight chamber was described in France as useful for reducing the mass and putrescible nature of suspended organic material from municipal wastewaters [40]. In 1897 the Local Government Board of the city of Exeter, England, approved the treatment of all city's wastewater in an anaerobic tank. Imhoff developed, at the beginning of this century in Germany, the Emscher tank to treat not only the wastewater but also the sludge. This tank has two chambers, the upper one for the wastewater and the lower one for the sludge (Fig. 7).

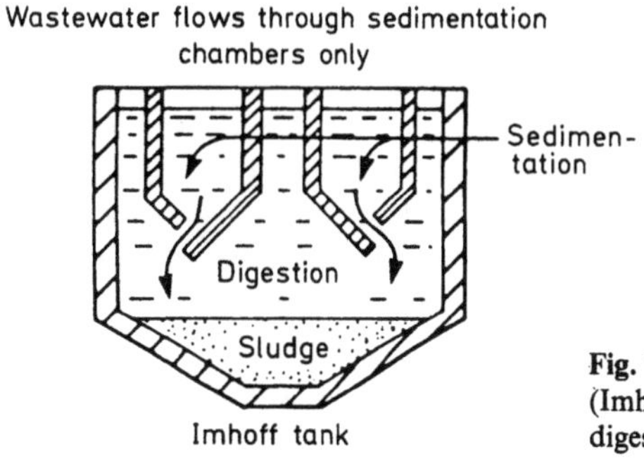

Fig. 7. Scheme of an early anaerobic treatment system (Imhoff tank) which combined sedimentation and digestion in a single unit [40]

The sludge was allowed to stay in this chamber for periods from a week to several months.

To optimize the anaerobic sludge treatment, attempts were made to carry out digestion in a tank separated from the sedimentation tank. In 1927 the Ruhrverband installed the first heating apparatus in a separate digestion tank, to use the methane produced by this process. Numerous studies led to better understanding of the anaerobic system. Thus by the end of 1930s the anaerobic treatment of municipal sludges in heated digesters was well established. During the 1950s it was found that mixing enhanced the rate of digestion by bringing bacteria and wastes more closely together and destroyed the scum layers.

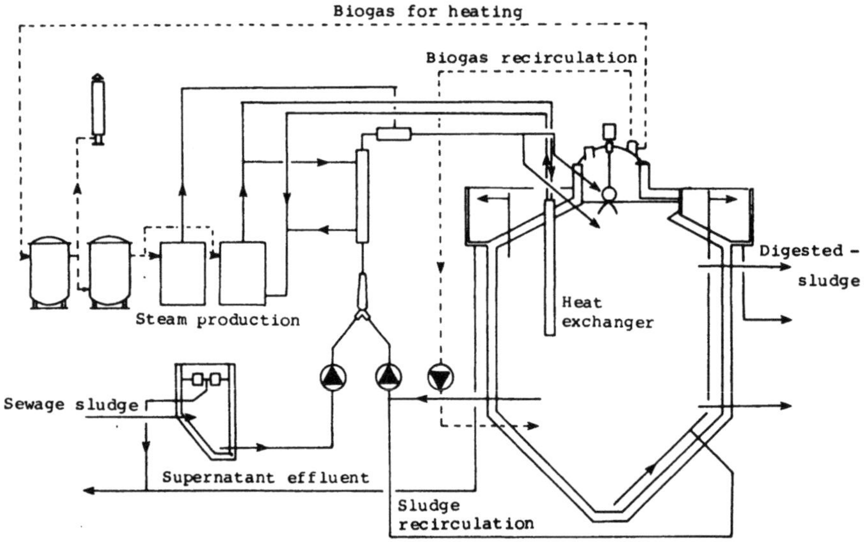

Fig. 8. Modern digestion system, commonly used for sewage sludge digestion (Roediger, Hanau)

Therefore modern digesters (Fig. 8) usually are mixed by sludge recirculation, gas recirculation, mechanical draft-tube mixers, or turbine and propeller mixers. They are heated to 30–35 °C either by recirculating the digester contents through an external heat exchanger or by using an internal heat exchanger through which the reactor contents move at high velocity to prevent fouling of the heat exchange surface. In this way the residence time of the sludge could be reduced from 30–40 d to about 15–20 d, and the volumetric load could be increased from 1 to 4 kg organic dry matter m^{-3} d^{-1}. Usually about 50–60% of the organic material is decomposed; this means that about 2–2.5 m^3 biogas are formed per day and per m^3 digester volume [41]. This biogas contains about 60–70% (v/v) methane and has therefore an caloric content of 5500 kcal m^{-3}. The digesters usually have capacities between 500 and 700 m^3; the largest European digester tanks, each with a reaction volume of 12000 m^3, have been set up at Düsseldorf in 1975 (Fig. 9).

Fig. 9. Digestion plant of the municipal sewage treatment plant at Düsseldorf

3.2 Anaerobic Contact Process

For the treatment of wastewaters from food and bioindustries, the anaerobic contact process has been developed. In an early work on this process Stander recognized in 1950 the value of maintaining a large population of bacteria in the methane-producing reactor [42]. The flow pattern is essentially the same as that used in the activated sludge process. A mean cell residence time significantly greater than hydraulic retention time is achieved by settling the solids from the effluent and recirculating a concentrated sludge back to the fermenter.

As shown in Fig. 10 this plant consists of a reaction tank followed by a settling tank (clarifier). The wastewater is treated in a continuously stirred tank reactor with an active population of flocculated bacteria degrading the organic material into methane and carbon dioxide. The effluent passes through a sludge settler where the flocculated bacteria settle to the bottom and are then returned to the reaction tank.

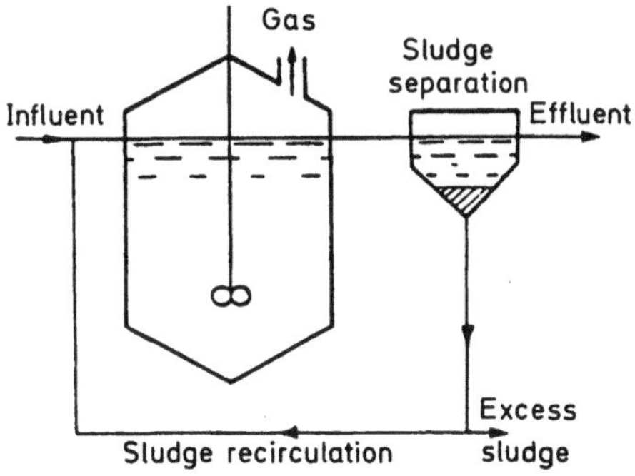

Fig. 10. Anaerobic contact process designed to maintain high concentration of bacteria at short hydraulic detention times

Because the bacteria are retained and recycled, this type of plant can treat medium strength wastewater (2000–20000 COD mg l^{-1}, very efficiently at high hydraulic loading rates.

A major difficulty encountered with this process is the poor settling properties of the anaerobic biomass from the effluent liquor. The methanogenic bacteria continue to produce biogas and the gas bubbles tend to adhere to the bacterial flocs. In order to overcome the poor sedimentation problem the following methods have been used or have been recommended: chemical flocculation, vacuum degasification, air stripping, flotation or centrifugation.

The Anamet system, an application of the anaerobic contact process, is successfully used for combined anaerobic-aerobic treatment of wastewaters from sugar factories and other food industries[43]. The first full scale plant was built at a sugar factory in Sweden 1972. It consists of a 20000 m^3 complete mixed anaerobic reactor followed by a lamella unit for sludge sedimentation and recycle. Using wastewater from beet sugar factories the average COD reduction in the anaerobic step was 90% at a feed rate of about 2–5 kg m^{-3} d^{-1} (COD). The gas production of CH_4 was 325 l NTP per kg COD added.

Recently at the central sludge treatment plant of the Emschergenossenschaft, West Germany, an anaerobic contact plant for the treatment of liquors from heat-treated sludge was constructed [44]. Three digester units work with an effective volume of 6000 m^3 each with a diameter of 25 m and an effective filling level of 12 m. The anaerobic activated sludge is retained within three gastight sedimentation tanks with a total volume of 1320 m^3. After approx. two years operation the following average values were obtained:

Load:	2273 m^3 d^{-1}
Digestion time:	7.9 d
Sludge content:	20 g l^{-1}
Organic sludge content:	10 g l^{-1}
COD in influent:	14.8 g l^{-1}
COD in effluent:	5.0 g l^{-1}
COD reduction:	67%
Biogas production:	9000 m^3 d^{-1}

The capital costs represent the main cost factor in this process. Since the investment costs for the entire plant were about 8.75 million DM, the capital costs amount to about 875000 DM per annum. In comparison the operating costs, 335100 DM annum, are considerably lower. Referring these capital and operating costs to the amount of COD reduction, it means that the decomposition of 1 kg COD costs about 0.15 DM. This is comparable to the costs for aerobic waste water treatment. However, in the case of the anaerobic contact process the biogas produced daily is equal to about 5300 l of oil. With a fuel oil price of 0.60 DM per l, this represents a saving of 1.16 million DM per annum if the biogas can be used. This is about the sum total of the capital and operational costs and thus this process is economically interesting.

3.3 Anaerobic Filter and Fluidized Bed Reactor

In order to overcome the difficulty in recycling the bacteria in the anaerobic contact process Young and McCarty [45] developed the anaerobic filter concept, which is similar to the aerobic trickling filter process. As shown schematically in Fig. 11 this process consists of a reactor filled with an inert support material such as gravel, rocks, coke or some plastic media. The packing should have a high surface onto which the microorganisms can attach. Therefore, in this case there is no need for biomass separation and sludge recycling. Long sludge ages at very high hydraulic loading rates are possible. Once an active microbial culture is established, the anaerobic filter demonstrates a remarkable resilience to variable loading rates and to moderate environmental changes such as pH or temperature. The population of microorganisms that develops in the reactor is in equilibrium with the organic loading and the hydraulic retention time. Since this is an upflow column, suspended matter accumulates in the column.

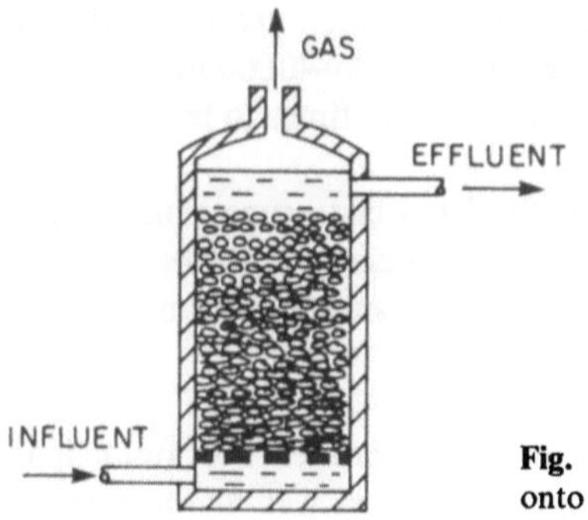

Fig. 11. Anaerobic filter containing inert support material onto which the microorganisms are attached [40)]

If this material is biodegradable, an equilibrium will be established at which the rate of deposition equals the rate of biodegradation. However, refractory suspended solids will accumulate. If the content of these solids in the wastewater is high, this accumulation may be a problem, requiring frequent blowdown of the column. Furthermore, because of the obvious danger of these filters becoming blocked with excess biomass their use is usually restricted to wastewaters with COD concentrations between 1000–10000 mg.

This anaerobic filter saw its first full-scale application in 1972 in the treatment of wheat starch wastewater [46)]. For carbohydrate-based wastewaters this process readily achieves 80–90% removal of COD at space-loadings of 2–4 kg m^{-3} d^{-1} COD. The most impressive performance is for purification of wastewaters containing acetic acid, methanol and formic acid which are direct substrates for the methane bacteria and which have very low biomass yield coefficients. Here the anaerobic filter's outstanding ability to retain biomass permits COD removals of 90% at a space loading of 10–20 kg m^{-3} d^{-1} COD with virtually no surplus sludge production.

Another recent development by van den Berg and coworkers [47)] is a process similar to that developed by McCarty, the so called thin film reactor. In this system the acetogenic and methanogenic bacteria adhere to the inside surface of tubes made from red draintile clay. Each reactor contains several tubes which have an inside diameter of 5–10 cm and a length of about 140 cm (Fig. 12). By admitting wastewater from above, rather than below as is done with anaerobic filters, problems of channelling and plugging are avoided. The film formed on clay was about 1–3 mm thick and had an activity of about 0.8–1.2 g^{-1} d^{-1} COD. Maximum methane production rates of 5 m^3 d^{-1} were achieved in a reactor with 100 m^2 film support area per m^3 from bean blanching wastewater.

This fixed biofilm concept was extended by developping the fluidized bed reactor. This process consists of passing the aqueous stream to be treated upwards through a bed of fine particles, such as sand (diameter 0.2–1 mm), at sufficient velocity to expand the bed beyond the point at which the frictional drag is equal to the net downward force exerted by gravity. Recirculation of the wastewater is usually necessary to maintain a high flow rate (Fig. 13). This fluidization allows the entire surface area of each grain to become available for biological colonization. Bacteria grow as a film on the surface of these particles and degrade the organic material. Surface areas on the order of 300 m^2 m^{-3} of bed are common in the fluid-bed system, resulting in extremely high biomass concentrations which are generally an order of magnitude greater than conventional processes (800–40000 mg volatile suspended solids l^{-1}). This allows the reactor size and treatment times to be markedly reduced.

Factors that contribute to the effectiveness of the fluidized bed process include:

1. Maximum contact between the liquid and the fine particles carrying the bacteria.
2. Problems of channeling, plugging, and gas hold-up commonly encountered in packed bed are avoided.
3. Ability to control and optimize biological film thickness.

The process has been successfully employed in the treatment of municipal wastewater [48], and industrial effluents [49] for nitrogen control as well as the aerobic BOD treatment. Pilot-scale testing has shown that anaerobic fluidized bed reactors

Fig. 12. Thin film reactor in which the methanogenic bacteria adhere to the inside surface of the tubes

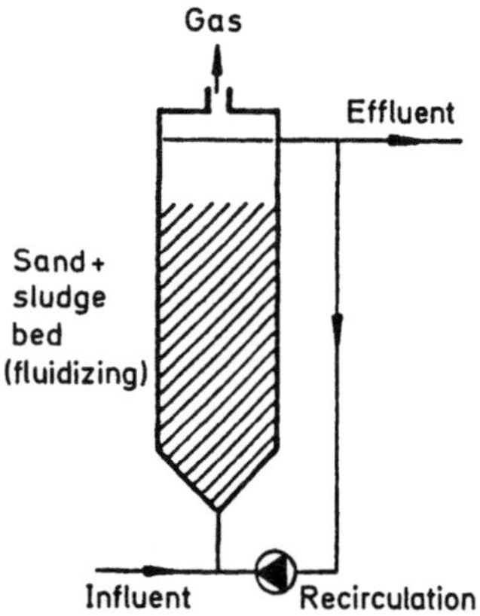

Fig. 13. Fluidized bed reactor in which the wastewater is passed upwards through a bed of particles carrying the bacteria

are very effective for the treatment of various wastes including dairy, chemical, food processing, soft drink bottling and heat treatment liquors[50]. In most cases, COD can be reduced by more than 80% at organic loading rates of 16 kg m^{-3} d^{-1} COD or higher. Large swings in organic loading rates and temperature may be tolerated without significantly deteriorating the treatment efficiency. Both high- and low-strength wastewaters originating from diverse sources can be treated successfully. Attachment of the bacteria to inert, fast-settling particles permits very good solid/liquid separation and offers the possibility of using the anaerobic digestion process to treat very dilute wastewaters (100–1000 mg COD per 1), e.g. municipal sewage. A full scale anaerobic fluidized-bed treatment plant is now installed at a bottling plant because the economics of the system appear quite favorable [50].

3.4 Upflow Anaerobic Sludge Blanket (UASB) Reactor

The upflow sludge blanket process for anaerobic wastewater treatment has been developped in the Netherlands. It was first used to treat maize-starch wastewaters in South Africa [51], but its full potential was only realized after an impressive programme of development by Lettinga in the 1970s [52]. As shown in Fig. 14 the simple design of the UASB reactor is based upon superior settling properties of the sludge. The wastewater is fed into the UASB digester from below and leaves at the top via an internal baffle system for separation of the gas, sludge and liquid. With this device gas is separated from the sludge and it is collected underneath the plates and piped away. Above the plates a relatively quiet zone is created where the sludge is separated from the fluid and can settle back towards the digesting zone. With this method adequate sludge residence times and high anaerobic sludge concentrations in the digesters can be obtained. Since forced mechanical stirring will have a detrimental effect on the settling characteristics of the sludge, the reactor is mainly mixed by the gas production. It was found that 80–90% of the substrate degradation occurred in the lower part of the reactor where a very high concentration of active anaerobic sludge is present. As there is no mechanical mixing a sludge gradient over the height of the tank is formed, characterized by a concentration

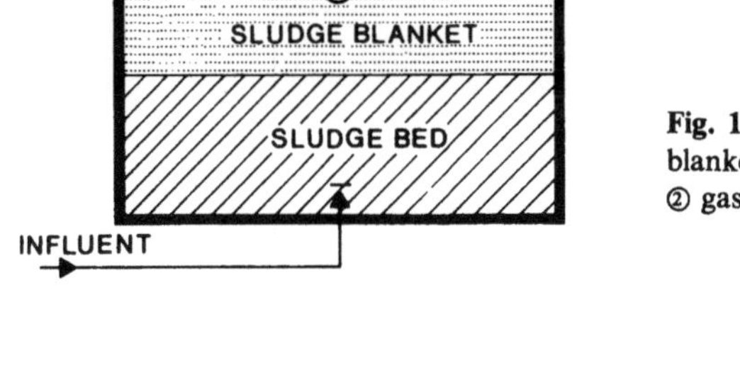

Fig. 14. Scheme of an upflow anaerobic sludge blanket reactor. ① Sludge-liquid mixture inlet, ② gasscreens, ③ settled sludge return opening [55]

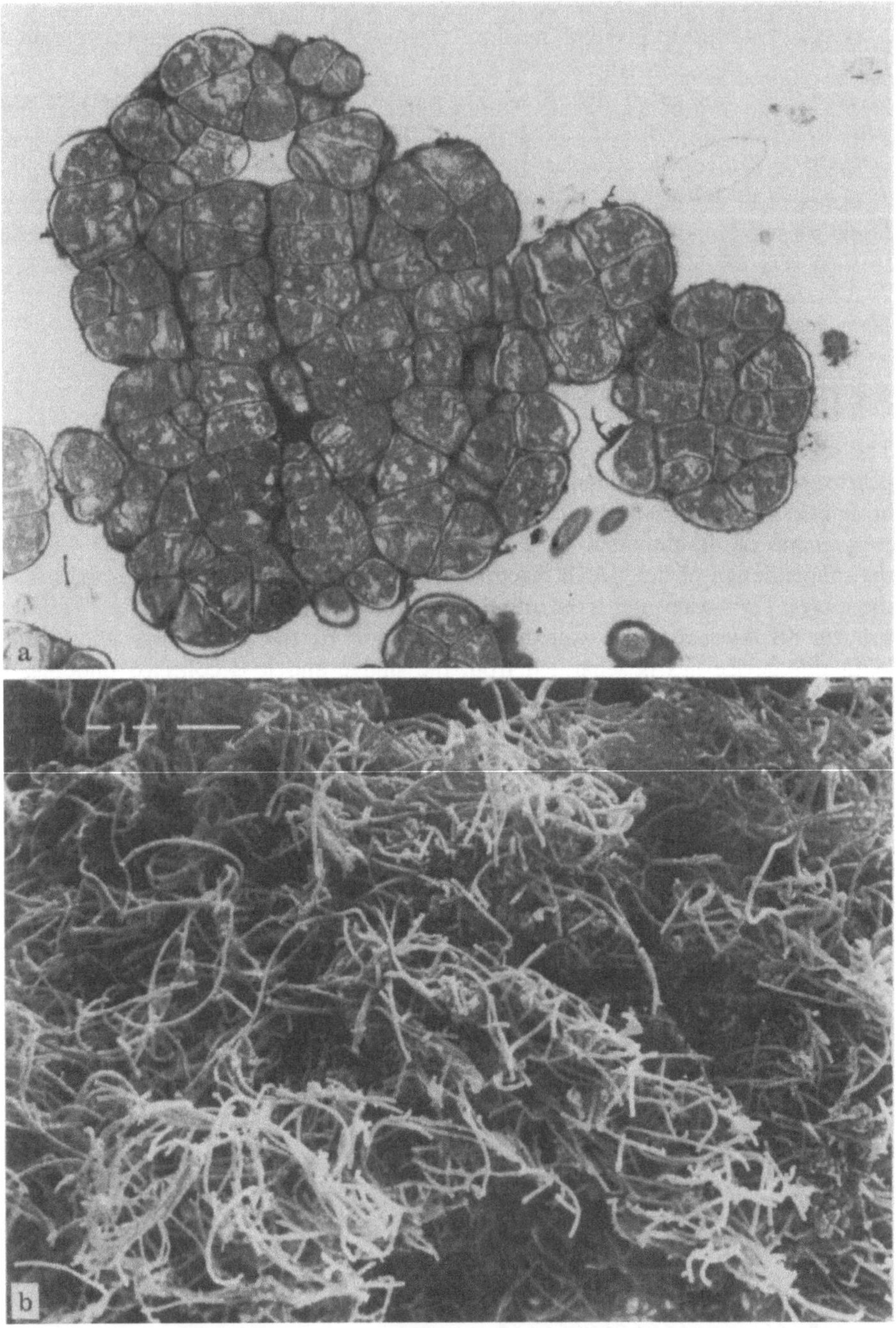

Fig. 15a. Typical cell aggregates of *Methanosarcina* sp. obtained by high stationary concentrations of acetic acid. **b** Typical thin filaments enriched at low stationary concentrations of acetic acid [53]

of 50 to 100 kg m^{-3} of sludge near the bottom and 5 to 20 kg m^3 above it. The height of the sludge bed should be 1.5–2.5 m; at heights of 0.4 m most of the influent bypasses the bed. The process accomodates fairly well to hydraulic and organic shock loads, temperature fluctuations, and low influent pH values, provided the digester pH remains well above pH 6.0 and that the sludge load applied is below the maximum specific COD removal rate. Under optimal conditions a COD-loading of 15 kg m^{-3} of reactor volume per day and more could be treated with an efficiency of 70–90%. Hydraulic retention times of as low as 4 h were feasible, with an excellent settling sludge.

Undoubtely one of the main features of the UASB process is the development of a granular sludge, which is highly settleable and can be mixed well by the circulating gas. These granules have a diameter of 0.5 to 2.5 mm and possess excellent settling properties 12×10^{-3} m s^{-1}. Until now the pellet formation is not sufficiently understood, but some factors have been recognized to be important in this process:

1. The presence of sufficient nurtrients for bacterial growth and for the formation of the bonding agents. Filamentous bacteria, which may cause bulking sludge, mainly occur during periods of substrate limitation. For example methanogenic sarcina-forming cell clusters were most abundant in digesters fed with high acetate concentrations. However, at low steady state acetic acid concentrations thin and slender filaments were enriched (Fig. 15) [53].
2. The continuous washout of non- or poorly flocculating bacteria. One of the main merits of the UASB-concept is that it selects for the best settling sludge particles.
3. Calcium ions have an evident positive effect on the flocculating ability of anaerobic sludge.

Table 3. Upflow anaerobic sludge blanket treatment plants so far buildt [55]

Year of realisation	Industry	Country	Volume m^3	Capacity kg d^{-1} COD
1976	Liquid sugar	Netherland	30	500
1977	Beet sugar	Netherland	200	3000
1978	Beet sugar	Netherland	800	13000
1979	Beet sugar	Netherland	1424	20700
1980	Potato-proc.	Netherland	240	1600
	Potato-proc.	Netherland	400	3200
	Potato-starch	Netherland	1700	13540
	Beet-sugar	Netherland	1300	16200
	Potato-proc.	Switzerland	600	5000
1981	Brewery	USA	4600	66000
	Potato-proc.	Netherland	1500	16000
	Beet-sugar	Germany	1500	18000
	Beet-sugar	Netherland	1700	29000
	Beet-sugar	Netherland	1800	28000
	Alcohol	Netherland	700	11000
	Candy	Netherland	100	1040
	Potato-starch	USA	1800	20000
1982	Beet-sugar	Austria	3040	25000

4. Intensive mechanical mixing should be prevented, because it may promote the disturbance of the granules of the sludge.
5. The initial sludge load should be below 0.1–0.2 kg kg^{-1} COD total solids per day and the loading rate should not be increased unless all volatile acids are decomposed.

Until now the UASB reactor has been applied mostly in the beet sugar industry (Table 3) [55]. Scaling-up of upflow reactors in horizontal directions is no problem if the influent distribution has been given special attention. With the present knowledge the optimum reactor height seems to lie between 4 and 6 m [54]. Operational results from three plants in Holland are given in Table 4. The average hydraulic retention time is 4 to 6 h at a COD loading of 12–16 kg m^{-3} d^{-1} with a COD-reduction of about 70–75%. The average biogas amounts to 0.42 m^3 kg^{-1} COD removed. These high loading rates are possible because of the high sludge concentration in the reactor.

Table 4. Operational results of three UASB plants at sugar industries [55]

	Factory: Halfweg	Breda	Groningen
Reactor volume (m^3)	800	1300	1425
Hydraulic flow (m^3 h^{-1})	200	275	250
HRT (h)	4	4.8	5.7
COD infl. (ppm)	1850	2400	4000
COD load (kg m^3 d^{-1})	12	12	16.5
COD reduction (%)			
total	70	75	75
soluble	85	85	85
Gas production (m^3 h^{-1})	120	200	300
Gas composition (% CH_4)	82	82	76

In the potato-processing industry some difficulties were experienced at first to obtain good sludge settling characteristics. For that reason, it was decided to change the operation of the plant into a twostep process whereupon COD loading rates of 5 kg m^{-3} d^{-1} were feasible. From these results it can be concluded that the UASB process is a very suitable system for the treatment of industrial wastewaters.

Obviously, there is no single process design that is ideal for all kinds of wastewaters. Improved process designs are continually being developed and reactors are becoming more stable and economic. Methane fermentation has traditionally a bad reputation for process instability. If, howeverer, the solids retention time can be made sufficiently long (> 100 d) process stability is rather good. Data for some anaerobic processes are shown in Table 5.

4 Environmental Factors

To achieve the desired result in anaerobic wastewater treatment, the environmental requirements of the different microorganisms cooperating in the methane process must be fulfilled. The methanogenesis is mainly influenced by the following three parameters:

Table 5. Characteristics of anaerobic treatment processes [65)]

Parameter	High rate digester	Contact process	Anaerobic filter/fluidized bed process	UASB-process
Suitable wastes and wastewaters	Sludge, manure, solid wastes	Wastewaters from food industry	Wastewaters from food industry	Sugar industry wastewaters
Suitable COD-concentration, mg l^{-1}	>20.000	2.000–20.000	500–10.000 (40.000)	1.000–50.000
Treatment temperature 0 °C	a) 30–40 b) 55–60	30–40	20–35	7–35
Organic load COD kg m^{-3} d^{-1}	1–8	1–5	1–15	3–15
Sludge load, kg COD per kg sludge	–	0.2– 0.5	–	0.8–1.0
Waste sludge production kg sludge per kg COD degraded	–	0.03–0.10	0.03–0.10	0.04
Solids retention time, d	a) 10–30 b) 5–15	>20	>100	>100
Hydraulic retention time, d	a) 10–30 b) 5–15	0.5–25	0.2–3	0.2–3
Effiency, % COD	30–70	60–90	70–95	80–90
Methane content of gas, % (v v^{-1})	50–75	50–90	50–90	80–90

1. Composition of wastewater
2. Temperature
3. PH and volatile acids

4.1 Composition of Wastewater

Recently, it was demonstrated that anaerobic bacteria do not only degrade carbohydrates, proteins and lipids quite well but also some petrochemicals as shown in Table 6 [56]. Cultures that can degrade benzoic acid have been enriched from sewage digesters, mud and the rumen of cattle [16,57]. The stoichiometry of methanogenesis from benzoic acid follows [9]:

$$4\,C_6H_5COOH + 18\,H_2O \rightarrow 15\,CH_4 + 13\,CO_2\,. \qquad (9)$$

More complex aromatic compounds can also be degraded to methane. For example Healy and Young [58] have demonstrated methane formation from vanillin, ferulic acid, syringic acid, phenol and 4-hydroxybenzoic acid. Studies on the anaerobic metabolism of these compounds indicate that the aromatic ring is reduced prior to the cleavage, which results in the formation of aliphatic acids (Fig. 16).

Table 6. Petrochemicals metabolized by enriched methane cultures [56]

Acetaldehyde	Formic acid	Propanal
Acetone	Fumaric acid	Propanol
Adipic acid	Glutaric acid	2-Propanol
1-Amino-2-propanol	Glycerol	Propionic acid
4-Amino-butyric acid	Hexanoic acid	Propylene glycol
Benzoic acid	Hydroquinone	Resorcinol
Butanol	Isobutyric acid	Sec-butanol
Butyraldehyde	Maleic acid	Sec-butylamine
Butyric acid	Methanol	Sorbic acid
Catechol	Methyl acetate	Succinic acid
Crotonaldehyde	Methyl ethyl ketone	Tert-butanol
Crotonic acid	Nitrobenzene	Valeric acid
Ethyl acetate	Phenol	Vinyl acetate
Ethyl acrylate	Phthalic acid	

Furthermore an enrichment culture was obtained, which was able to convert furfural completely into methane and CO_2 [53]. As shown in Fig. 17 acetic acid was detected as an intermediate in the conversion of furfural into biogas. The strain which degrades furfural into acetate belongs to *Desulfovibrio* species.

At present it seems that only very few organic compounds cannot be decomposed by anaerobic microorganisms. These are lignin, n-paraffins, some plastics and components with ether linkages; these substances seem to need oxygen for their microbial decomposition.

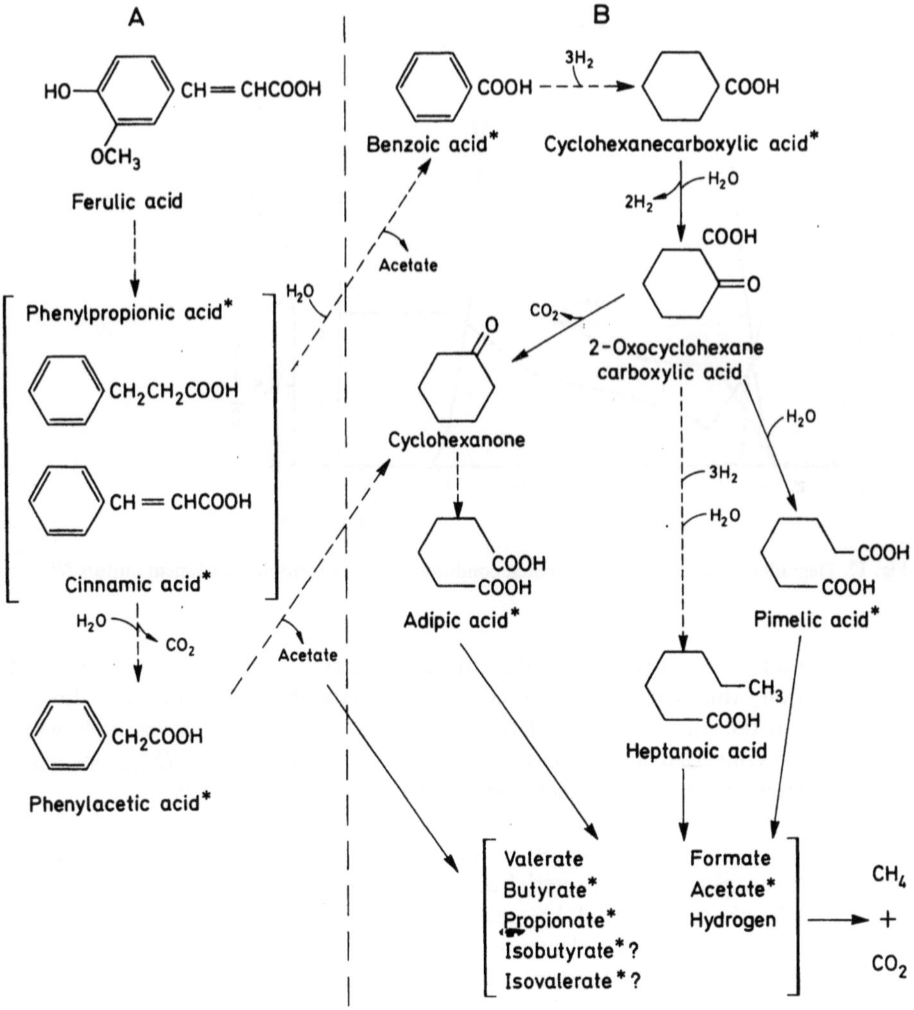

Fig. 16. Suggested model for decomposition of ferulic acid to methane [58]

Organic matter destruction is directly related to methane production, from 1 kg COD degraded about 350 l methane are formed. Buswell and Mueller [59] developed Eq. (10) to calculate the produced amount of methane and CO_2 in the biogas from a knowledge of the chemical composition of the degraded waste.

$$C_nH_aO_b + \left(n - \frac{a}{4} - \frac{b}{2}\right) H_2O \rightarrow \left(\frac{n}{2} - \frac{a}{8} + \frac{b}{4}\right) CO_2 + \left(\frac{n}{2} + \frac{a}{8} - \frac{b}{4}\right) CH_4 \qquad (10)$$

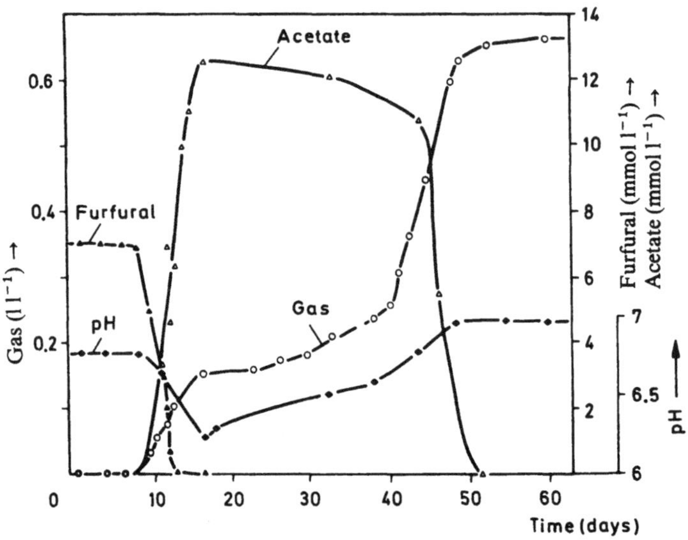

Fig. 17. Degradation of furfural under batch conditions by an anaerobic enrichment culture [53]

This equation demonstrates that the organic material is degraded by a kind of disproportion; this means that the content of methane in the biogas is directly correlated to the oxidation step of the organic waste material. For example when alcohols are converted into biogas the methane content is about 75%, but when carbohydrates are utilized the methane content in the biogas is only 50%.

For growth, the bacteria require besides the carbon and energy source also some inorganic salts for the synthesis of cell material. Besides 54% carbon, 20% oxygen and 10% hydrogen the dry bacterial cell mass contains on an average about 12% nitrogen, 2% phosphorus and 1% sulfur, as well as sodium, potassium, calcium, magnesium and several trace elements such as iron, manganese, molybdenum, zinc, copper, cobalt, selenium, tungston, nickel etc. Scherer et al. [60] demonstrated that the growth of the methanogenic bacterium, *Methanosarcina barkeri*, is dependent on cobalt and molybdenum. With 10^{-6} M Co and 5×10^{-7} M Mo the growth is optimal (Fig. 18). For optimal growth the cells require Co to build up the Co containing corrinoid Factor III (0.1–0.2 mg 5-hydroxylbenzimidazolylcyanocobamide per g wet cells). Furthermore, Schönheit et al. [61] found that the growth of *Methanobacterium thermoautotrophicum* depends on nickel. The formation of 1 g dry weight cells required approximately 150 nmol nickel. An extension of the work to other organisms revealed that nickel is generally required for methanogenic bacteria [62]. These organisms contain a nickel tetrapyrrole cofactor, F_{430}, which is involved in methane formation [63,64].

While the municipal wastewaters contain usually all of these components in sufficient amounts, many industrial wastewaters are not nutritionally balanced and often lack nitrogen, phosphate and trace elements. In such cases it is necessary to add the components in sufficient amounts to the wastewater before treatment. However, the nutrient requirements of the anaerobic bacteria are much lower than of the aerobic

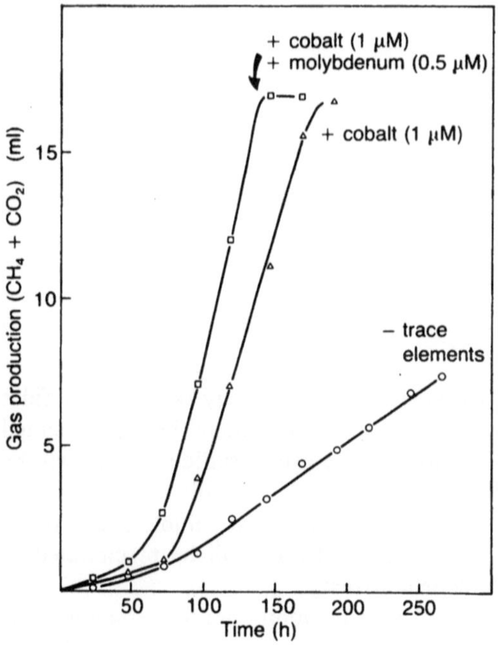

Fig. 18. Effect of cobalt and molybdenum on methanogenesis of *Methanosarcina barkeri*[60]

ones, since only a small amount of bacteria biomass is formed. Requirements for nitrogen and phosphorus in anaerobic treatment processes were calculated and determined to be in the C:N:P ratio of 700:5:1[65].

Several investigatiors have observed that sulfate inhibits methanogenesis in digesters[66]. The possible reasons for this effect are:

1. Competition for substrates; the available data suggest that the sulfate reducing bacteria are able to out-compete the methanogens for immediate substrates. This would correlate with the fact that slightly more free energy is released during sulfate-reduction than during the reduction of carbon dioxide to methane[11].
2. Inhibition of the methanogens by sulfide, which is formed during bacterial sulfate reduction. Sulfate itself is not appreciably toxic to isolated methanogens[67].
3. Precipitation of essential trace elements (e.g. Fe, Ni, Co, Mo) by sulfide formed.

Sulfide inhibits methane formation in digesters at 1–6 mmol l^{-1}[66,68]. It is not clear whether these inhibitory effects are the result of competition for hydrogen by sulfate-reducing bacteria or of direct inhibition of methanogenic bacteria. It seems that only very low stationary sulfide concentrations are necessary for the growth of methanogens in pure culture[30,69].

Heavy metals are toxic to the anaerobic population at very low concentrations as shown in Table 7[70]. However, since the toxicity only seems to be concerned with the free metal ions, the toxicity depends very much on complexing or precipitating anions. In this respect the formation of sulfides is particulary important because heavy metal sulfides are extremely unsoluble, their solubility product ranges from 3.7×10^{-19} for FeS to 8.5×10^{-45} for CuS. Therefore, when the digester feed contains sufficient sulfur compounds (0.5 mg mg^{-1} sulfide heavy metal), fairly high concentrations of

Table 7. Concentrations of soluble heavy metals exhibiting 50% inhibition of anaerobic digesters [70]

Cation	Approximate concentration in mg l^{-1}
Fe^{++}	1–10
Zn^{++}	10^{-4}
Cd^{++}	10^{-7}
Cu^{+}	10^{-12}
Cu^{++}	10^{-16}

heavy metals are admissible in the wastewater. If the naturally occurring sulfides are not sufficient to prevent heavy metal toxicity, then they should be supplemented by the addition of ferrous sulfate [71]. In this way the excess sulfide formed will be held out of solution by the iron. If additional heavy metals enter the reactor they will draw the sulfide from the iron since iron sulfide is the most soluble heavy metal sulfide. As long as the pH is above 6.4 the iron released will be precipitated as iron carbonate, thereby preventing soluble iron toxicity.

Besides sulfate and heavy metals the list of possible inhibitors of methanogenesis includes chlorinated hydrocarbons, detergents and antibiotics. Chloroform and other halogenated methane analogues were found to be very potent inhibitors of methanogenic bacteria in sewage digesters at a concentration of approximately 1 ppm [72]. Likewise fluoroacetate inhibits the formation of methane from acetate [73] and bromoethansulphonic acid, an analogue of coenzyme M, has been shown to block methane formation [74]. Detergents at a concentration of about 15 mg l^{-1} have caused difficulties in sewage-digesters. When a sudden spillage or overuse of detergent occurs, a small digester may be inhibited because it does not have sufficient sludge to dilute the toxic material, whereas a larger digester would be able to withstand the shock. The antibiotic monensin, which is used in the animal feed as an additive, causes a reduction of methanogenesis already at a concentration of 1 µg mol^{-1} [75].

In order to prevent failure of the anaerobic system it is necessary to identify inhibition of methanogenesis at an erly stage. Parameters commonly used as inhibition indicators are:

1. Reduction in methane yield, which should normally be in the range of 0.34 to 0.36 m^3 CH_4 per kg COD removed at 35 °C or 0.91–0.93 m^3 CH_4 per kg organic carbon removed.
2. Increase in volatile acids concentration, which should normally be less than 250 mg l^{-1} within the reactor. Volatile acid concentrations above 500 mg l^{-1} indicate either that the organic loading rate is too high or that the system is inhibited. A trend of rising propionic acid concentration is a good indicator that the acetogenic bacteria have been inhibited [76].

4.2 Temperature

Microbial formation of methane has been demonstrated at temperatures between 0 °C (glacier ice) and 97 °C (hot springs [77]). There seem to be two temperature ranges within this wider range; these correspond to two sets of bacteria, the mesophiles which operate best at 5–40 °C with an optimum growth temperature between 20 and 40 °C, and the thermophiles with an optimum between 50 and 75 °C.

In Fig. 19 with the temperature dependence of methane formation for different strains of methanogenic bacteria, the two temperature optima are quite obvious. Besides the thermophilic strains: *Methanobacterium thermoautotrophicum* and *Methanococcus thermolithotrophicus*, an extreme thermophile, *Methanothermus fervidus* has recently been isolated from a hot spring in Iceland [78]. This strain grows between 63 °C and 97 °C with an optimum of 83 °C. So far only one slightly thermophilic acetate utilizing methanogen has been described: *Methanosarcina* strain TM-1 [79], but it can be expected that others occur since acetate is quite well degraded into biogas at 60 °C [53]. All the other methane bacteria described up to now are typical mesophiles. Although methane is also formed at 4 °C psychrophilic methane bacteria have not yet been isolated.

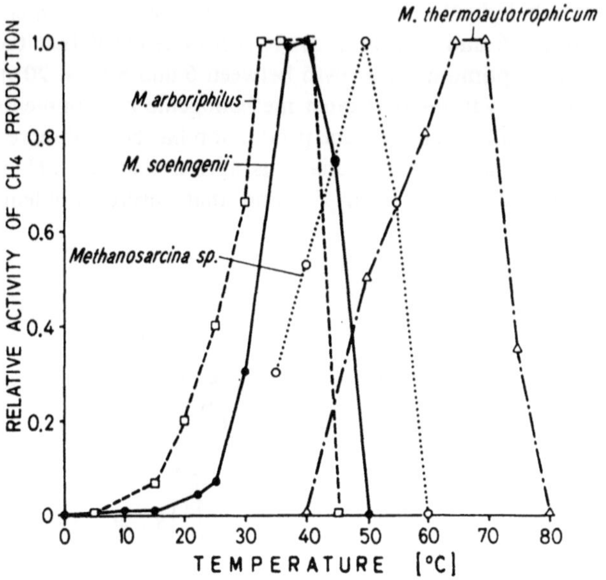

Fig. 19. Temperature relationship for methane formation by different methanogenic bacteria [77]

In spite of the fact that greater gas production can be expected if a digester is operated in the thermophilic range, this is very rarely done because the energy required to maintain the digester at a suitable temperature more than outweighs the increased gas production. Also thermophilic bacteria are rather more sensitive to changes in environmental conditions than mesophilic ones, so that the degree of control incorporated in the design would make a thermophilic digester costly. It might be used, however, when the incoming waste is already at a high temperature, as in an industrial context.

Anaerobic digesters at municipal sewage treatment plants are usually operated at temperatures between 30 and 35 °C. This temperature range combines the best conditions for the growth of bacteria and for the production of methane with the shortest retention time of the waste in the digester. For wastewaters where solids retention is possible, lower treatment temperatures may be used, because mesophilic sludge still excerts an appreciable activity at temperatures as low as 10 °C [80].

When the anaerobic system is in operation, the temperature should be kept as constant as possible. Mosey [81] stated that temperature should not fluctuate by more than 2 °C per day. The need for a constant temperature can be explained by the different behaviours of the three groups of bacteria. Owing to their faster growth rates, the acidogens adapt more rapidly to changed conditions than methanogens, causing an accumulation of metabolic products (organic acids). This results in an overall imbalance which can lead to process failure. Consequently, a uniform temperature is more important than a temperature giving maximum rates.

4.3 pH and Volatile Acids

Practical experience and research indicated early that anaerobic waste treatment is optimal in a pH range between 6 and 8 (Fig. 20a). It is now realized that the methanogenic bacteria have a pH-optimum for growth between 6 and 8 (Fig. 20b). This is in accordance with the observations that most methanogenic environments function optimally at a neutral pH, although methane production has been observed in peat at a pH of 4 [82]. For methanogenic bacteria it is agreed that the pH of anaerobic operations should be maintained near 7.0 and that severe problems can result if the pH drops below 6.

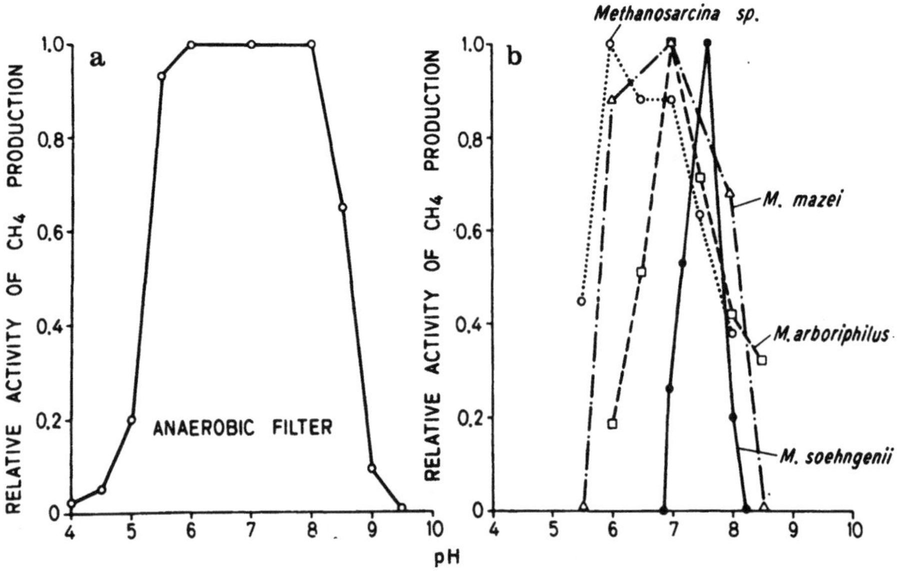

Fig. 20. PH relationship for methane formation in an anaerobic filter (**a**) and with various methanogenic bacteria (**b**) [77]

Under balanced digestion conditions, the biochemical reactions tend to maintain automatically the pH in the proper range. Although volatile organic acids produced by the acidogenic bacteria during decomposition of complex organisms tend to reduce the pH, this effect is counteracted by degradation of these acids and reformation of bicarbonate buffer during methane fermentation. However, if an imbalance of the process occurs, for example by a rapid change of the environmental conditions, the acid forming bacteria outpace the methanogens and organic acids build up in the system. If balanced digestion is not restored, the buffer capacity is overcome, and a precipitate drop in pH occurs. The low pH will stop methane production almost completely but will hinder acid formation only slightly; thus a „sour" digester results. Restoration of balanced conditions in a „sour" digester requires considerable time because of the low growth rate of methanogenic bacteria. In many cases, balance is achieved only by seeding heavily with anaerobic sludge.

Therefore, a prime objective in the operation of anaerobic wastewater treatment processes is maintenance of a proper pH range. The signal for imminent trouble is a sudden rise in the volatile acids. When the volatile acids rise suddenly the balance is restored by reducing the feed rate of the digester for several days. Another method is to raise the pH by adding alkaline substances such as calcium hydroxide. However, pH control is not a universal remedy; its only advantage is to prevent a bad situation from getting out of hand. The basic cause of the digester biochemical imbalance must be discovered and corrected. Unless this is done, pH control is worthless in the long run. In addition, care must be exercised in selecting an alkaline material that will not produce a toxic reaction.

The concentrations of volatile acids and alkalinity during anaerobic treatment depend on the concentration of the wastewater and on its composition. In a more dilute wastewater, volatile acids and alkalinity are removed at a relatively high rate with the effluent. Therefore the ratio of volatile acids to alkalinity is a better criterion for the stability of the system than the absolute values of these parameters. A ratio of total volatile acids (as acetic acid) to total alkalinity (as calcium carbonate) of less than 0.1 is desirable [83].

In order to stabilize the total process and to optimize the partial processes, a spatial separation of the acidogenic and methanogenic populations was proposed [84,85]. The two-phase digestion involves two biologically active digesters in series, which function to optimize conditions for active metabolism of microorganisms. In the first phase, complex organic substances (carbohydrates, proteins, fats) are hydrolyzed and fermented mainly to volatile fatty acids by anaerobic acidogens. During the second phase, the organic acids are converted to biogas by the acetogenic and methanogenic bacteria. This separation of the anaerobic process into two steps offers, as compared with the conventional process (one reactor), the following advantages [86]:

1. The conversion rates can be increased by optimizing the process conditions for both steps. Cohen et al. [87] found that maximum specific sludge loadings of the methanogenic phase of the two-phase system was over 3 times higher than that of the one-phase system.
2. This process is more stable, overloading the methane reactor can be prevented by proper control of the acidification step [88].

3. Compounds that are toxic to methane bacteria may be removed in the first reactor (e.g. sulfide).
4. The acidification reactor can serve as a buffer system; this is very useful when the composition of the wastewater is changed. By this means it is possible to give the slowly growing methanogenic bacteria more time for adaptation to the new conditions.

However, a disadvantage of this two-phase process is the more complex design (two reactors) and the necessity for more extensive measuring and control equipment. This system has been proved successful for anaerobic treatment of wastewaters from the sugar industry and from distilleries [89].

The choice between a one or two step system depends very much on the composition of the wastewater. For example, in pilot plant experiments it was demonstrated that liquid-sugar or potato-processing wastes can be purified equally well in a single reactor or in a two-phase digestor. Apparently in a well adapted system the acidogens and the methanogens work fairly well together in one reactor. Furthermore, if a system is operated under a low load, it is generally quite possible to use a one step system. However, there may be some wastewaters for which a two-phase digestion is very useful, e.g. wastes with a high content of undissolved matter. More research is necessary to answer the question whether the two stage process will be preferred over the one stage process.

5 Concluding Remarks

Methane fermentation is well known as important for the decomposition of organic materials under anaerobic conditions. This process was recognized in the 18th century; but the biochemical basis and the complex symbiotic relationships between the numerous bacterial species involved has only recently been elucidated. It could be described that at least three groups of bacteria are involved in the degradation of organic compounds into methane and CO_2, namely: the hydrolytic and acidogenic bacteria, the acetogens and finally the methanogens. Yet, a fundamental understanding is far from complete. Therefore, more research on the following areas can be of great benefit to anaerobic processes:

1. Characterization of the different bacterial strains which are involved in the degradation of various organic materials.
2. Studies on the relationship among the different microorganisms in mixed cultures and on the conditions for the formation of pellets [90].
3. Identification of rate limiting biochemical steps in the anaerobic decomposition of specific substrates.
4. Development of active starter cultures for specific wastewater treatment containing defined metabolic associations of anaerobic bacteria.
5. Genetic improvement of key microbial species in anaerobic digestion by selection and development of hyperactive strains.

The traditional application of methane fermentation is to sludge stabilization at municipal sewage treatment plants. Primary and excess sludge are anaerobically treated in conventional digestion tanks. Because of the increasing costs of energy this process has gained increasing interest in the treatment of wastewaters. Since some of the anaerobic bacteria have a very long generation time (several days), special reactors have to be developed to retain most of the microorganisms inside the digesters or to recycle the bacteria after separation. In this way the solids retention time is uncoupled from the fluid retention time and high bacterial concentrations are obtained in the digesters which give high degradation rates (COD > 15 kg m^{-3} d^{-1}). If the solid retention time can be made sufficiently long, process stability is rather good and the treatment of low strength wastewater (municipal sewage) even at low temperature (5–10 °C) seems possible. Furthermore recent results indicate that not only wastewaters from the food and bioindustries, but also many wastewaters from the pharmaceutical and chemical industries are amenable to anaerobic treatment. The development of anaerobic processes can be enhanced by major research efforts within the following areas:

1. Process development; although some of the modern processes are already applied at full scale, considerable improvements can be made in the various systems. Because of the varied wastes and treament needs, modifications are necessary to improve performance capabilities for given situations.
2. The influence of environmental factors like geometry, hydraulics and micronutrients on the optimization of reaction rates should be studied in more detail.
3. The mechanism of bacterial attachment to solid surfaces is of particular interest, furthermore, cheap carrier materials should be developed and investigated.
4. To increase the reliability of the process more control systems with very good stabilities over long time are necessary.

Thus there are many opportunities for innovation in anaerobic treatment, a process which offers an enormous potential for the removal of organic pollutants from different wastewaters while at the same time producing methane.

6 Acknowledgement

I am grateful to Professor R. K. Finn for his advice and correction of the English translation.

7 References

1. Zeikus, J. G.: Bacteriol. Rev. *41*, 514 (1977)
2. Barker, H. A.: Bacterial fermentations. New York: Wiley 1956
3. Ehhalt, D. H. et al.: Pageoph. *116*, 452 (1978)
4. Higgings, I. J. et al.: Microbial Rev. *45*, 556 (1981)
5. Hobson, P. N. et al.: Crit. Rev. Environ. Control *4*, 131 (1974)
6. Hungate, R. E.: The rumen and its microbes. New York: Acad. Press 1966
7. Bryant, M. P. et al.: Arch. Microbiol. *59*, 20 (1967)
8. Bryant, M. P. et al.: Appl. Environ. Microbiol. *33*, 1162 (1977)
9. Toerien, D. F. et al.: Water Res. *1*, 397 (1967)
10. Khan, A. W.: Can. J. Microbiol. *23*, 1700 (1977)
11. Thauer, R. K. et al.: Bacteriol Rev. *41*, 100 (1977)
12. Gottschalk, G.: Bacterial metabolism. New York: Springer 1979
13. Scheifinger, C. C. et al.: Appl. Microbiol. *26*, 789 (1973)
14. Bryant, M. P.: J. Animal Science *48*, 193 (1979)
15. McInerney, M. J. et al.: Arch. Microbiol. *122*, 129 (1979)

16. Ferry, J. G. et al.: Arch. Microbiol. *107*, 33 (1976)
17. Reddy, C. A. et al.: J. Bacteriol. *109*, 539 (1972)
18. Lawrence, A. W. et al.: J. Water Poll. Control Fed. *41*, 21 (1969)
19. McInerney, M. J. et al.: Abstr. Proc. Amer. Soc. Mibrobiol., *182*, 94 (1978)
20. Boone, D. R. et al.: Appl. Environ. Microbiol. *40*, 626 (1980)
21. McInerney, M. J. et al.: Appl. Environ. Microbiol. *41*, 1029 (1981)
22. Bonch-Osmolovskaya, E. A.: Mikrobiologiya *47*, 1014 (1978)
23. Hungate, R. E.: Methods in Microbiology *3B*, 117 (1969)
24. Zehnder, A. J. B. et al.: Arch. Microbiol. *111*, 199 (1977)
25. Balch, W. E. et al.: Microbiol. Rev. *43*, 260 (1979)
26. Taylor, G. T.: Prog. Indust. Microbiol. *16*, 231 (1982)
27. Hungate, R. E. et al.: J. Bacteriol. *102*, 389 (1970)
28. Huser, B. A. et al.: Arch. Microbiol. *132*, 1 (1982)
29. Smith M. R. et al.: Proc. Biochem. *15*, 34 (1980)
30. Scherer, P. et al.: Eur. J. Appl. Microbiol. Biotechnol. *12*, 28 (1981)
31. Hippe, H. et al.: Proc. Natl. Acad. Sci. (USA) *76*, 494 (1979)
32. Kandler, O.: Zbl. Bakt. Hyg. I. Abt. Orig. C *3*, 149 (1982)
33. Langworthy, T. A. et al.: Zbl. Bakt. Hyg., I. Abt. Orig. C *3*, 228 (1982)
34. Vogels, G. D. et al.: Zbl. Bakt. Hyg. I., Abt. Orig. C *3*, 258 (1982)
35. Binot, R. A. et al.: Biotechnol. Lett. *3*, 623 (1981)
36. Thauer, R. K.: Zbl. Bakt. Hyg. I., Abt. Orig. C *3*, 265 (1982)
37. Woese, C. R.: Zbl. Bakt. Hyg. I., Abt. Orig. C *3*, 1 (1982)
38. Daniels, L. et al.: J. Bacteriol. *136*, 75 (1978)
39. Fuchs, G. et al.: Zbl. Bakt. Hyg. I., Abt. Orig. C *3*, 277 (1982)
40. McCarty, P. L.: In: Anaerobic Digestion 1981 (ed. Hughes, D. E. et al.) p. 3. Amsterdam: Elsevier Biomedical 1982
41. Roediger, H.: Die anaerobe alkalische Schlammfaulung. Oldenbourg, München 1967
42. Stander, G. J.: J. Int. Sewage Purification *4*, 438 (1950)
43. Huss, L.: In: Anaerobic Digestion 1981 (ed. Hughes, D. E. et al.). p. 137. Amsterdam: Elsevier Biomedical 1982
44. Schlegel, S. et al.: In: Anaerobic Digestion 1981 (ed. Hughes, D. E. et al.) p. 169. Amsterdam Elsevier Biomedical 1982
45. Young, J. C. et al.: J. Water Pollut. Control Fed. *41*, R 160 (1969)
46. Taylor, D. W.: Proc. 3rd Nat. Symp. Food Processing Wastes p. 151 (1972)
47. van den Berg, L. et al.: Biotechnol. Lett. *3*, 165 (1981)
48. Jeris, J. S. et al.: J. Water Pollut. Control Fed. *49*, 816 (1977)
49. Hancher, C. W. et al.: Biotech. Bioeng. Symp. *8*, 361 (1978)
50. Hickey, R. F. et al.: Biotech. Bioeng. Symp. *11*, 399 (1981)
51. Hemens, J. et al.: Wat. Waste Treat. *9*, 16 (1962)
52. Lettinga, G. et al.: Biotech. Bioeng. *22*, 699 (1980)
53. Brune, G. et al.: Proc. Biochem. May/June *20* (1982)
54. Heertjes et al.: Biotech. Bioeng. *24*, 443 (1982)
55. Pette, K. C. et al.: In: Anaerobic Digestion 1981 (ed. Hughes, D. E. et al.), p. 121. Amsterdam: Elsevier Biomedical 1982
56. Chou, W. L. et al.: Biotech. Bioeng. Symp. *8*, 391 (1978)
57. Evans, W. C.: Nature *270*, 17 (1977)
58. Healy, J. B. et al.: Appl. Environ. Microbiol. *39*, 436 (1980)
59. Buswell, A. M. et al.: Ind. Eng. Chem. *44*, 550 (1952)
60. Scherer, P. et al.: Acta Biotechnologica *1*, 57 (1981)
61. Schönheit, P. et al.: Arch. Microbiol. *123*, 105 (1979)
62. Diekert, G. et al.: J. Bacteriol. *148*, 459 (1981)
63. Pfaltz et al.: Helv. Chim. Acta *65*, 828 (1982)
64. Keltjens, J.: PhD-Thesis University Nijmegen 1982
65. Frostell, B.: PhD-thesis University Stockholm 1979
66. Khan, A. W. et al.: Appl. Environ. Microbiol. *35*, 1027 (1978)
67. Patel, G. B. et al.: J. Appl. Bacteriol. *45*, 347 (1978)
68. Lawrence, A. W. et al.: Air Water Pollut. Int. J. *10*, 207 (1966)
69. Cappenberg, Th. E.: Microbial Ecol. *2*, 60 (1975)

70. Mosey, F. E. et al.: Water Pollut. Control *74*, 18 (1975)
71. Lawrence, A. W. et al.: J. Water Pollut. Control Fed. *37*, 392 (1965)
72. Thiel, P. G.: Water Res. *3*, 215 (1969)
73. Winfrey, M. R. et al.: Appl. Environ. Microbiol. *37*, 244 (1979)
74. Gunsalus, R. P. et al.: Biochem. *17*, 2374 (1978)
75. Hilpert, R. et al.: Zbl. Bakt. Hyg., I. Abt. Orig. C *2*, 11 (1981)
76. Anderson, G. K. et al.: Proc. Biochem. July/August 28 (1982)
77. Zehnder, A. J. B. et al.: In: Anaerobic Digestion 1981 (ed. Hughes D. E. et al.) p. 45. Amsterdam: Elsevier Biomedical 1982
78. Stetter, K. O. et al.: Zbl. Bakt. Hyg. I. Abt. Orig. C *2*, 166 (1981)
79. Zinder, S. H. et al.: Appl. Environ. Microbiol. *38*, 996 (1979)
80. Lettinga, G. et al.: In: Anaerobic Digestion 1981 (ed. Hughes, D. E. et al.). Amsterdam: Elsevier Biomedical 1982
81. Mosey, F. E.: Symp. Treatment of Wastes from Food and Drink Industry (Newcastle) p. 1 (1974)
82. Svensson, B. H.: In: Production and Utilization of Gases (ed. Schlegel, H. G. et al.) p. 135 Göttigen 1976
83. Kaspar, H. F. et al.: Microbiol. Ecology *4*, 241 (1978)
84. Pohland, F. G. et al.: Environ. Lett. *1*, 255 (1971)
85. Ghosh, S. et al.: Proc. Biochem. April, *15* (1978)
86. Zoetemeyer, R. J.: PhD-thesis University Amsterdam 1982
87. Cohen, A. et al.: Water Research *14*, 1439 (1980)
88. Cohen, A.: PhD-thesis University Amsterdam 1982
89. Pipyn, P. et al.: Biotechnol. Lett. *1*, 495 (1979)
90. Bochem, H. P. et al.: Can. J. Microbiol. *28*, 500 (1982)

Role and Function of Protozoa in the Biological Treatment of Polluted Waters

Ryuichi Sudo
Laboratory of Freshwater Environment, National Institute for Environmental Studies. P. O. Tsukuba-Gakuen, Tsukuba, Ibaraki, 305 Japan

Shuichi Aiba
Department of Fermentation Technology, Faculty of Engineering, Osaka University, Yamada-oka, Suita-shi, Osaka 565 Japan

The activated sludge used in the biological treatment of polluted water is a typical ecosystem composed of bacteria and protozoa. There should be an indigenous picture of competition, predation, mutualism and so forth, among the microorganisms which contribute to the removal of organic matters from polluted waters.

Specific protozoa have been detected in the sludge when the process proceeds satisfactorily. Therefore, a study how to control the protozoan fauna in the sludge to warrant a good quality of effluent is essential.

This article presents a clue leading to the microbiological control of the biotic community in the sludge, through the studies on ecological characteristics in the monoxenic culture of ciliated protozoa isolated from the activated sludge.

1 Quick Review of Activated Sludge and Bio-film Processes

Whatever processes are used for biologically treating polluted waters, it is evident from an ecological standpoint that its typical ecosystem comprizes protozoa, bacteria, fungi and metazoa.

Interactions such as competition, commensalism, mutualism (symbiosis), synergism and predation occur amongst the microorganisms in the ecosystem and as a consequence, BOD, COD or other nutrients are removed from polluted waters. Accordingly, the ecological viewpoint requires to develop the continuous biological treatment that produces an effluent of excellent quality. Several auxiliary procedures for solid separation in grit chambers and in primary sedimentation tanks must precede the biological oxidation system, which must be followed by the secondary sedimentation tank and chlorination regardless of the kind of biological process employed.

1.1 Flow Diagram of Activated Sludge Process

A flow diagram of the activated sludge process is shown in Fig. 1. The preliminary treatment of polluted waters involves the removal of suspended materials by physical means, i.e. large solids are first removed from the water by passage through metal-bar screens and grit chambers with linear velocities of about 0.3 m s^{-1}. Then the polluted water passes to the primary sedimentation tank, where at retention times of between 1 to 3 h, some 60% of the remaining total suspended solids equivalent to about 40% of the BOD, is removed efficiently. The supernatant liquid and the sediment in the primary sedimentation tank are called "settled sewage" and "settled sludge", respectively.

The settled sludge is usually blended with excess sludge that comes from the final sedimentation tank and anaerobically digested.

The settled sewage is aerated for 3 to 8 h in the aeration tank, to which a significant fraction of the activated sludge from the secondary sedimentation tank is recycled. After the biological oxidation of pollutants has occurred, the mixed liquor suspended solid (MLSS) in the aeration tank passes to the secondary sedimentation tank where it is divided into a clarified supernatant and a concentrated activated sludge stream by sedimentation. Most of the concentrated activated sludge is recycled to the aeration tank (as return sludge), whilst the excess sludge proceeds to sludge digestion.

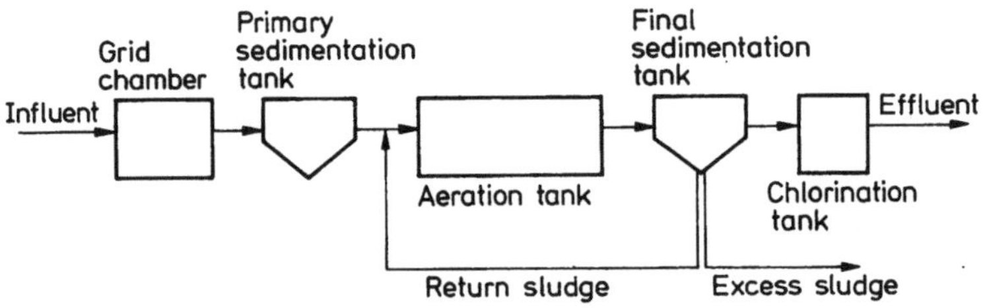

Fig. 1. Flow diagram of activated sludge process

The secondary treatment of wastewaters comprises biological oxidation and sedimentation. In recent years, tertiary treatment to remove phosphorus and nitrogen to a far greater extent than is achieved in secondary treatment has been advocated to retard the eutrophication in terrestrial and littoral waters.

1.2 Flow Diagram of Bio-film Process

A flow diagram of the bio-film process is shown in Fig. 2. The oxidation is performed by a bio-film attached to either a solid packing material or rotating discs that are either dosed with or partially submerged in wastewaters, respectively.
The principal characteristic of the bio-film process is in the recycling of clarified supernatant from the secondary sedimentation tank to the bio-film rather than recycling sludge as in the activated sludge process. The process operates with biomass retention rather than biomass recycle.

In the activated sludge process the microbial habitat forms disposed flocs; in the bio-film process it comprizes a film on the surface of solid support material or rotating discs. In the latter case, the discs usually made of plastic, 1 to 5 m in diameter, are mounted on a horizontal shaft passing through an oxidation tank and are rotated slowly (about 1 to 5 rpm), keeping approximately 40% of the surface area immersed in the tank. During rotation, a liquid film of polluted water forms on the disc surface as it is alternately exposed to the air and immersed in the liquid. This alternate exposure and immersion allows microbial growth as a bio-film on the disc surface. This process has been introduced for small-scale wastewater treatment.

The trickling filter comprizes either a bed of crushed rocks or plastic packing, and the wastewater is sprinkled over the bed. Trickling filters are still extensively used for municipal wastewaters, but at present the activated sludge process is usually preferred.

Fig. 3 shows a schematic diagram of the metabolism in bio-film processes. Usually, the bio-film is bushy and matted with many macroscopic filaments which project outward into the adjacent film of polluted water. This provides an active biological surface area much larger than just that of the support media. It also enables nutrients and dissolved oxygen to reach a greater portion of the biomass and render it aerobically active. The thickness of the bio-film which permits the penetration of oxygen is usually 2 to 3 mm. Accordingly, when BOD loading is high and the thickness of bio-film is more than 3 mm, the outer portion of the bio-film quickly consumes most of the available oxygen and the inner side becomes anaerobic.

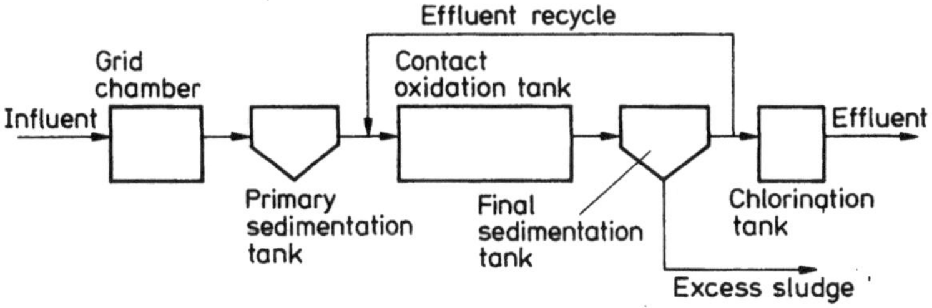

Fig. 2. Flow diagram of bio-film process (bio-film process is classified, according to the mode of contact oxidation, into submerged bio-film, rotating biological discs, and trickling filter)

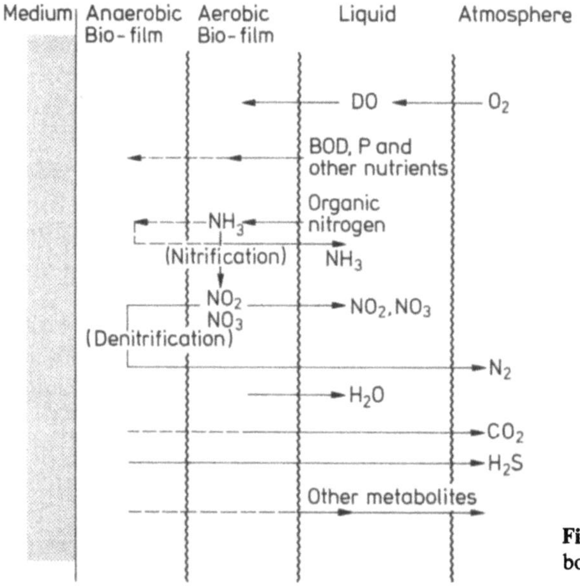

Fig. 3. Schematic diagram of metabolism in bio-film

The anaerobic area inside the bio-film is considered to cause periodically dislodging of the bio-film from the supporter surface, a process known as sloughing pieces. In the anaerobic zone, offensive odors from H_2S, volatile fatty acids and amines are generated. It is now recognized that a moderate sloughing of the anaerobic zone of the bio-film, by either biological or physical means, avoids the unfavorable effects of this zone.

Although the differences in both construction and operation of the activated sludge and the bio-film processes are obvious, the microbial spectrum extending from bacteria to metazoa, and the microbial role in removing pollutants from wastewaters are definetely shared by both processes, when certain trivial differences in microbial activity in the respective processes are overlooked.

The particular advantage of both processes is that they are cost effective for BOD removal (more than 90%), but unfortunately deficient to remove nitrogen and phosphorous. Biological processes are highly effective for the removal of biologically degradable organic substances when compared with physico-chemical processes.

2 Protozoan Fauna

Good quality effluents can be produced by activated sludge and bio-film type processes despite differences in their microbial flora and fauna. In this section emphasis will be placed on the protozoan fauna appearing in both biological processes to see whether difference in and/or the dominance of specific protozoan species could affect the effluent quality.

Photomicrographs of activated sludge and bio-film are shown in Figs. 4 and 5. Protozoa such as *Vorticella microstoma* are clearly evident in both figures. Apparently,

Fig. 4. Activated sludge biota (magnification × 150) of domestic wastewater treatment. See *Vorticella convallaria*

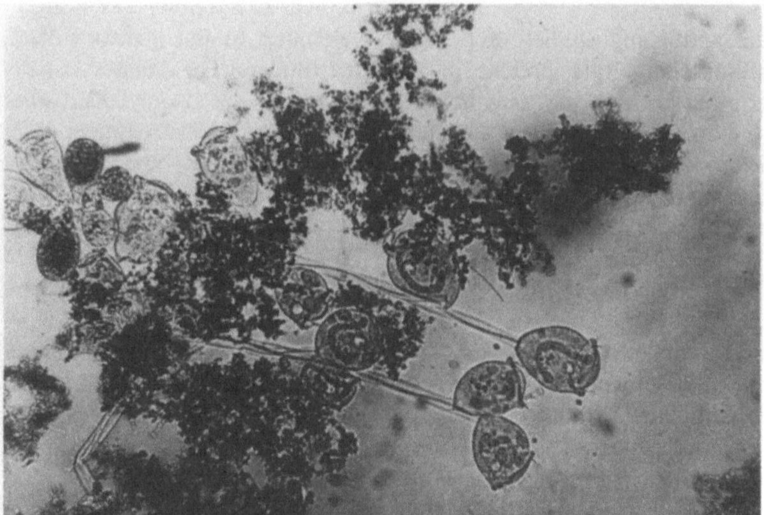

Fig. 5. Bio-film biota (magnification × 150) of rotating biological discs for domestic wastewater treatment. See *Carchesium polypinum* and *Tokophrya* sp. around bacterial flocs

the floc of the activated sludge, whose constituents are mainly bacteria and protozoa, is slack compared to the bio-film. In the latter, filamentous microorganisms and metazoa such as Rotifera and small aquatic Oligochaetes are observed in addition to bacteria and protozoa. Consequently, the bio-film is more closely packed.

The biofilm could be characterized by inhabitants of many smaller metazoa

Table 1. Protozoa emerging frequently in the biological wastewater treatment

A) Activated sludge	B) Bio-film
1. *Vorticella microstoma*	1. *Epistylis* sp.
2. *Vorticella convallaria*	2. *Vorticella convallaria*
3. *Aspidisca costata*	3. *Operncularia* sp.
4. *Epistylis plicatilis*	4. *Euglypha* sp.
5. *Vorticella campanula*	5. *Arcella vulgaris*
6. *Opercularia* sp.	6. *Vorticella microstoma*
7. *Vorticella alba*	7. *Cinetochilum margaritaceum*
8. *Aspidisca lynceus*	8. *Vorticella* sp.
9. *Opercularia coaractata*	9. *Carchesium polypinum*
10. *Carchesium polypinum*	10. *Zoothamnium* sp.

which occupy the upper trophic level in the food chain. However, since protozoa are shared by both activated sludge and bio-films, the principal species are compared in Table 1 without elaboration of the smaller metazoa that specifically inhabit the bio-film.

2.1 Protozoa Found in Activated Sludge

Large numbers and many species of protozoa belonging to the genera Ciliata, Flagellata and Sarcodina are present in activated sludge. The ciliates usually dominate over other protozoa not only in number of species but also in total count. A survey on the protozoa residing in the sludge from a wastewater treatment plant in Japan disclosed Ciliata (37 families), Flagellata (10 families) and Sarcodina (3 families)[1]. Of these protozoa, stalked Ciliata such as *Vorticella, Carchesium, Zoothamnium, Opercularia* and *Epistylis* predominated in normal activated sludge at all times.

Protozoa found in activated sludge when the effluent was of good quality are shown in Table 2. In this example, the aeration tank was operated at a MLSS 1,000–2,000 mg l^{-1} and a BOD loading, 0.2–0.3 kg kg $MLSS^{-1}$ d^{-1}. The population number in Table 2 pertains to the number of protozoan cells per ml of mixed liquor. As is apparent in Table 2, the predominant species of protozoa in the activated sludge were *Vorticella, Epistylis* and *Aspidisca.*

2.2 Protozoa Found in the Bio-film

Despite the differences in the structure of bio-films in trickling filters, submerged filters and rotating discs, some species of protozoa are common to all systems. Although protozoan species found in bio-films are similar to those found in activated sludge, the population of protozoa in the bio-film as a fraction of the whole population, is far larger than in the activated sludge. Branching stalked Ciliata such as *Carchesium, Epistylis, Opercularia* and *Zoothamnium* tended to become predominant.

These branching stalked Ciliata occasionally aggregate into large colonies of several hundred cells. The population density of protozoa is usually scored either

Table 2. Example of protozoa and its population in the activated sludge when good-quality effluent is obtained

Treatment plant MLSS (mg l^{-1})	A 1,190	B 1,470	C 1,960
Vorticella nebulifera	360	220	
Vorticella convallaria	1,260	3,720	
Vorticella longifillum	60	120	
Vorticella microstoma		60	3,140
Epistylis plicatilis	620	120	
Epistylis sp.		1,000	420
Zoothamnium arbuscula	920		200
Opercularia sp.			640
Aspidisca lynceus	2,920		
Aspidisca sp.	160		
Euplotes sp.	20		
Carchesium polypinum	200		720
Loxophilum spp.	300	140	
Litonotus sp.	420	460	850
Amphileptus spp.	40		
Astasia sp.	60		
Dileptus sp.		20	
Tetrahymena sp.		20	
Oikomonas termo	60		
Peranema sp.	20		
Bode edax			100
Tokophrya spp.	60	180	
Acineta spp.		80	
Podophrya sp.		20	
Amoeba spp.		280	40

Note: Numerical value is number ml^{-1} of activated sludge (mixed liquor)

by the number per cm^2 of surface area of support media or mg of dry weight of bio-film.

Examples of protozoa found in rotating biological discs are shown in Table 3. Although the protozoa listed are not substantially different from those found in activated sludge, the protozoan population mg^{-1} in the first stage of rotating biological discs is far larger than in activated sludge processes i.e., the population of Ciliata such as *Epistylis*, *Opercularia* and *Vorticella* is usually 2,000–5,000 cells mg^{-1} (10,000–25,000 cm^{-2}). In the third and fourth stages, Sarcodina such as *Euglypha*, *Arcella* and *Amoeba* appear in larger numbers than the Ciliata.

2.3 Comments on the Role of Protozoa

Generally, a microscopic examination of both the protozoan species and population in sludge, regardless of its origin, can be a useful criterion in assessing any anomaly in process operation and/or in judging the quality of effluent from the treatment process[2,3].

Table 3. Example of protozoa and its population in the rotating biological discs

Biota	Stage 1	2	3	4
Amount of bio-film (mg cm^{-2})	5.10	4.25	1.80	1.15
Epistylis sp.	2,500	800		
Vorticella convallaria	1,000	600		
Cinetochilum margaritaceum	100	650		
Zoothamnium aselli		200		
Coleps hirtus				250
Opercularia sp.	1,000	800	200	
Uroleptus musculus				300
Colpoda cucullus				200
Euglypha sp.			200	1,600
Arcella vulgaris			400	1,000
Bodo sp.	150	600	700	
Peranema trichophorum				250
Pleuromonas jaculans			1,200	3,000
Amoeba sp.			1,200	500
Philodina roseola			420	60
Dorylaimus sp.	80			
Chaetogaster sp.	30	350	300	900
Nais sp.			200	

Note: Numerical value is population number mg^{-1} of bio-film

a. Protozoan candidates for assessing the performance of the activated sludge are:
   1. Protozoa usually found when the activated sludge is normal; *Vorticella*, *Epistylis*, *Aspidisca*, *Opercularia*, *Zoothamnium*, *Carchesium*, *Euplotes*, *Tokophrya*, *Podophrya* and *Acineta*.
   2. Protozoa frequently found when an anomaly occurs; *Bodo*, *Cercobodo*, *Oikomonas*, *Paramecium*, *Vahlkampfia*, *Metopus* and *Caenomorpha*.
   3. Protozoan species detected when the operations is between stage 1 and 2; *Litonotous*, *Loxophilum*, *Chilodonella*, *Oxytricha* and *Amoeba*.

b. Protozoan species to estimate the bio-film process are:
   1. Protozoan species present when the normal operation occurs; *Carchesium*, *Zoothamnium*, *Epistylis*, *Opercuralia*, *Vorticella*, *Aspidisca*, *Oxytricha*, *Arcella*, *Euglypha* and *Euplotes*.
   2. Protozoa present when an anomaly occurs; *Oikomonas*, *Bodo*, *Cercobodo*, *Paramecium*, *Colpidium*, *Holophrya*, *Glaucoma*, *Metopus* and *Caenomorpha*.
   3. Protozoan species present when operation is between 1 and 2; *Pleuromonas*, *Cinetochilum*, *Trachelophyllum*, *Spirostomum*, *Amoeba*, *Litonotous*, *Loxophilum* and *Amphileptus*.

Taking for granted that these protozoa identifications could be applied for judging the process performance, the total protozoan population of the activated sludge was plotted against the quality of effluent as shown in Fig. 6[1,3]. Regardless of the

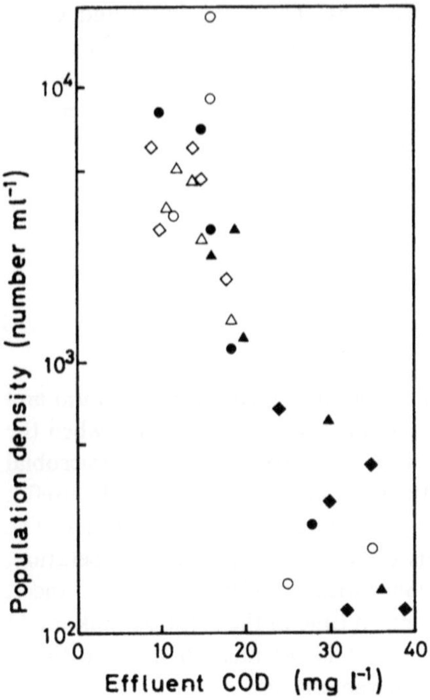

Fig. 6. Relationship between population number of activated-sludge protozoa and effluent COD [3] observed in sewage treatment plants in Tokyo Metropolitan area; ○ : Sunamachi, ▲ : Morigasaki, ● : Mikawashima, ◇ : Ochiai, △ : Shibaura, ◆ : Odai

wastewater treatment plant considered, it is clear from the figure that the effluent becomes excellent in quality with increasing protozoan population density with respect to the normal operation. In other words, effluent COD and suspended solids (SS) are low when the protozoan population is large, while the population of other protozoa in category 2 increases when the effluent COD and SS are high.

Similarly, the protozoan identification with respect to the bio-film provides additional means of assessing whether the bio-film either functions normally or abnormally. The most significant roles of protozoa in wastewater processes are

1) to separate bacteria efficiently from the mixed liquor during sedimentation by enhancing the flocculation of bacteria [2,3];
2) to remove dispersed bacteria by adsorbing them onto metabolite of protozoa and sinking rapidly with protozoa [3];
3) to decrease the concentration of dispersed bacteria because of the prey-predator relationship between protozoa and bacteria [2,4];
4) to provide bacteria with more "food of soluble substrate" by reducing the number of bacteria due to the predatory performance of protozoa [3];
5) to take up directly substrate [3];
6) to help "sanitize" the water by diminishing pathogenic microorganisms as well [2,3].

Effluents from wastewater treatment plants are required to be transparent, hygienically safe and free of soluble organic materials. The above-mentioned roles 1) to 3) contribute to a transparent effluent, i.e. to an increased transparency and a decreased concentration of suspended solids (SS), because SS in effluents are mainly microorganisms remaining unsedimented. The roles 4) to 5) contribute to decreased residual concentrations of soluble organic substances. Thus, the BOD and COD values might be reduced with both SS and soluble organic substances. The sixth role is important in achieving hygienic and safe effluents. In addition, relevant protozoa in the ecosystem reduce the amount of excess sludge significantly, because bacteria in the sludge are the substrate for the protozoa.

3 Characteristics of Protozoa

3.1 Food Habit [5]

Vorticella microstoma, Vorticella convallaria, Carchesium, Epistylis, Opercularia and *Aspidisca* can usually be isolated from the ecosystem in wastewater plants when the process is functioning satisfactorily. An idea of how to control the microbial flora and fauna in the process to acquire an effluent of good quality would be useful. Means of controlling the microbial biota can be divided into two categories. The first deals with physical and chemical controls of the community via temperature, pH, dissolved oxygen concentration and the concentration of any toxic substances. The intensity of aeration and the hydraulic retention time in the aeration tank might be used as parameters in this context. The second is oriented to the nutritional requirements of protozoa, i.e., their ecological characteristics.

Here, the protozoan nutritional requirements are called simply their "food habit". It is important to study the food habit of the biota step by step and to describe the purification of polluted waters. The first step is to clarify the food habit of a specific protozoon. Vorticellidae were selected as they represent the protozoan population in biological treatment when it functions effectively.

Notwithstanding the long history of the biological treatment of polluted waters, even the simple question of whether a dominant species of protozoa either metabolizes insoluble organic substances (bacteria) or directly utilizes soluble organic ingredients in the wastewater remains unanswered. In this respect, this is the first systematic study to determine not only the specific growth rate but also the food habit of *Vorticellidae*. Such a study might eventually permit the biochemical and microbiological control of a biological treatment process [5)], even though the protozoan population occupies only about 10% by weight of the community.

Values of specific growth rate, μ measured monoxenically at 20 C with representative protozoa isolated from the activated sludge at the Sewerage Bureau, Tokyo Metropolitan Government are summarized in Table 4 [6,7)]. The bacterial and heterogeneous food given to each protozoon was isolated from the activated sludge, in which the respective protozoa were dominant. The question of wheter the difference in μ values in the table came from either the characteristics of each species of protozoa or from the bacterial food supplied remains to be answered. However, the value of μ(= 1.4 per day) at 20 C for *Paramecium caudatum* examined as control was in good agreement with data presented by other workers [8)].

When the bacterial population was obtained from bulked sludge, wherein filamentous organisms dominate, and served as food for the Vorticellidae, the growth of *Vorticella microstoma* apparently deteriorated [5)]. This observation suggests specific

Table 4. Specific growth rate of protozoa at 20 C (protozoa isolated from the activated sludge were fed with the sludge bacteria)

Species	Growth rate		Ecological type
	t_d (h)	μ_p (d^{-1})	
Vorticella microstoma	5.0	3.3	Sessile
Vorticella convallaria	7.6	2.2	Sessile
Carchesium polypinum	9.3	1.8	Sessile
Opercularia sp.	5.0	3.3	Sessile
Epistylis plicatilis	10.2	1.6	Sessile
Aspidisca costata	13.6	1.2	Crawling
Aspidisca lynceus	12.4	1.3	Crawling
Colpidium campylum	4.7	3.6	Free swimming
Tetrahymena pyriformis	4.5	3.7	Free swimming
Paramaecium caudatum	12.0	1.4	Free swimming
Oikomonas termo	4.3	3.9	Free swimming

* doubling time = $0.69/\mu_p$

requirements of the protozoa for food, implying bacterial food that is most suited for the growth of particular protozoa.

The food habit of *Vorticella microstoma* at 20 C is summarized in Table 5. Among 15 species of pure baterial cultures examined, the four species listed at the upper part of the table were inadequate growth in monoxenic culture, i.e., doubling time = infinity. When *Flavobacterium suaveolens, Aerobacter aerogenes* and *Escherichia coli* were used as the food respectively, the protozoa could grow well regardless of the order, in which the food species were changed from one to another. However, the value of μ in each run was considerably smaller than for a mixed population of sludge bacteria used as control (cf. Table 4). Microscopic observations in these runs showed that the size of *Vorticella microstoma* and the number of food vacuoles were both reduced compared with those for control.

The remaining 8 species of bacteria in the table were confirmed to be potential protozoan food per se, although the values of μ were nearly two-thirds of those when the mixed bacterial population from the activated sludge was supplied as food. Except for the three species listed at the bottom of Table 5, the microscopic appearance of the protozoa was essentially similar to that when the sludge bacteria were supplied as food [5].

Table 5 demonstrates clearly that even bacteria in the same genera differed, depending on its species, in the appropriateness as a protozoan food. Similar experiments on *Vorticella convallaria* showed that the bacterial species suitable as food were rather limited. For instance, *Pseudomonas fluorescens* could serve as a food for *Vorticella microstoma*, even though the bacterium could not be utilized by *Vorticella convallaria. Carchesium polypinum*, a species of Vorticellidae could not grow monoxenically, even in a pure culture of bacteria listed in Table 5 [5].

It is most likely that the *Vorticellidae* studied here was stenophagic; i.e. the food habit of the protozoa was narrow and specific. However, among the three species of the Vorticellidae (*Vorticella microstoma, Vorticella convallaria* and *Car-*

Table 5. Growth rate of *Vorticella microstoma* (20 C) as affected by various kinds of bacterial food

Bacterial species as food of *Vorticella microstoma*	Growth rate of *Vorticella microstoma*	
	Doubling time t'_d (h)	Specific growth rate μ_p (d^{-1})
Pseudomonas ovalis IAM 1002	∞	0
Flavobacterium aquatile IFO 3772	∞	0
Achromobacter cycloclastes IAM 1013	∞	0
Micrococcus luteus IAM 1097	∞	0
Flavobacterium suaveolens IFO 3752	18.2	0.9
Aerobacter aerogenes IAM 1102	10.8	1.5
Escherichia coli IAM 1239	10.4	1.6
Alcaligenes faecalis IAM 1015	9.1	1.8
Bacillus subtilis IAM 1969	8.6	1.9
Bacillus cereus IAM 1029	7.9	2.1
Alcaligenes viscolatic IAM 1517	7.6	2.2
Pseudomonas fluorescens IFO 1039	7.6	2.2
Achromobacter liquidum IAM 1667	7.6	2.2
Micrococcus varians IAM 1313	7.6	2.2
Flavobacterium arborescens IFO 3750	7.6	2.2

chesium polypinum) examined, the bacterial species utilized as food for *Vorticella microstoma* were relatively extensive. This finding may substantiate the observation that among the Vorticellidae in the activated sludge, *Vorticella microstoma* frequently becomes dominant.

In connection with the food habit of the Vorticellidae, it is important to check whether these protozoa are holozoic, i.e., whether these protozoa metabolize only living bacteria or whether non-viable organic materials (soluble and/or insoluble in water) are also utilized.

A suspension of boiled egg yolk and either rice bran or suspended solids from a raw sewage in the BOD dilution water, containing 5,000 units ml^{-1} penicillin and 100 $\mu g\ ml^{-1}$ (streptomycin) could not sustain the growth of *Vorticella microstoma* indefinitely. On the other hand, Fineley et al. reported that an axenic culture of *Vorticella microstoma* could not be sustained, although the protozoa proliferated in the early period of a successive cultivation [9]. Similarly, the axenic culture of *Vorticella microstoma* in the sludge extract, containing 5,000 units ml^{-1} penicillin and 100 $\mu g\ ml^{-1}$ streptomycin was also unsucessful [5]. Judging from these results, it is most probable that the Vorticellidae are holozoic; the protozoa do not utilize organic materials directly irrespective of whatever they are either soluble or insoluble in water. The argument that the Vorticellidae are holozoic remains to be justified by further experimentation other than the above-mentioned observation.

The growth of *Vorticella microstoma*, when fed with a mixed population of bacteria as food, is shown in Table 6 [3]. Since the mixed food studied here was both restricted in bacterial species and arbitrary with respect to composition, it is difficult to draw conclusions on the effect of mixed bacterial populations on the growth of *Vorticella microstoma*. However, comparing Table 6 with Table 5 and calculating the growth rate of this protozoon fed with heterogeneous bacterial food isolated from the

Table 6. Growth rate of *Vorticella microstoma* as affected by various kinds of mixed baterial food [5] (continued)

Bacterial mixture as food of protozoa	Growth rate of protozoa	
	Doubling time t_d (h)*	Specific growth rate μ_p ($^{-1}$)
Pseudomonas fluorescens + *Flavobacterium arborescens*	6.8	2.4
Pseudomonas fluorescens + *Bacillus cereus*	7.1	2.3
Pseudomonas fluorescens + *Bacillus cereus* + *Flavobacterium arborescens*	6.7	2.5
Pseudomonas fluorescens + *Flavobacterium arborescens* + *Alcaligenes viscolactis*	6.3	2.6
Pseudomonas fluorescens + *Flavobacterium arborescens* + *Alcaligenes faecalis*	6.4	2.6

activated sludge as control (Table 4), the values of μ in Table 6 tend to approach those of the control.

Coler et al. [10] studied the food habit of *Colpoda* sp. by using 12 species of bacteria. Two categories of bacteria were defined as a result of their study; i.e. bacteria which apparently could be used as food by *Colpoda* (*Aerobacter aerogenes* and *Escherichia coli*), and bacteria to *Arthrobacter* which were ineffective in supporting protozoan growth.

Curds et al. [8], on the other hand, isolated 6 species of protozoa from activated sludge, and studied the protozoan food habit by employing 19 strains of bacteria. In their studies, bacteria were categorized as toxic, unfavorable and favorable as protozoan food. They concluded that the species of bacteria which were favorable as food varied, depending on the protozoan species. In another study, Fineley et al. [9] observed that *Bacillus cereus* was a favorable food of *Vorticella microstoma*. Their observations agree with the result shown in Table 5.

In light of the studies referred to above, it would be plausible to assume that changes of bacterial-flora in the aeration tank of activated sludge are closely related to those of the protozoan fauna. If a species of protozoa, *Vorticella microstoma* for instance, was needed for securing a good quality effluent, the above assumption would stress the importance of developing in the sludge the bacterial flora that is most suited as food for the protozoa. However, achieving a bacterial flora that is most desirable for a protozon is beyond our current level of knowledge and expertise. Ample room exists for further experimentation and discussion on the ways and means of controlling the emergence and/or disappearance of a protozoon in the sludge during biological wastewater treatment.

3.2 Growth Yield

Protozoa such as *Vorticella microstoma* have hardly ever been cultivated monoxenically, except on solid surfaces [5]. However, suspended, monoxenic cultures of stalked Ciliata are necessary to establish quantitative relationships between populations of protozoa and bacteria, i.e. prey-predator models. In this context, the monoxenic cultivation of stalked Ciliata in slowly shaken glass vessels was carried out successfully by Sudo and Aiba [11].

Following the previous work on the food habit of *Vorticella microstoma*, four strains of bacteria were used to study the mass cultivation of protozoae especially from the point of view of the feasibility of measuring the dry mass of single cells. *Alcaligenes faecalis* was the most favorable, because this bacterium formed only few bacteria-protozoa aggregates. If aggregates were formed, it was not possible to assess the dry mass of single cells of protozoa and bacteria, respectively.

The yield coefficient for the conversion of *Alcaligenes faecalis* to *Vorticella microstoma* was estimated as 0.47 g protozoa per g bacteria (Fig. 7) [11]. Since washed resting bacterial cells were used in this monoxenic cultivation for 72 h at 20 C, the spontaneous decrease in the mass of *Alcaligenes faecalis* due to autolysis was determined by cultivating the bacterial suspension, without protozoa to correct for the bacterial food consumed by the protozoa. This approach is valid to estimate the correct value of bacteria taken up by protozoa, i.e. the amount of bacterial autolysis is substracted from the total amount of decrease in bacterial population in the axenic culture.

Published data for protozoan yield from bacteria are shown in Table 7 [3,12–15]. In addition, other kinetic constants that have been obtained from batch culture are listed in Table 7. For details of the maximum growth rate, $\mu_{p,\,max}$ and saturation constant, K_x, see next section.

It can be noted from Table 7 that the yield for the conversion of bacteria to Ciliata is around 0.5, although the data of Proper and Garver [12], and Canale et al. [15] suggest a value considerably higher than 0.5. The yield coefficient depends on the accuracy of determining the dry weight of protozoa. The determination of the dry

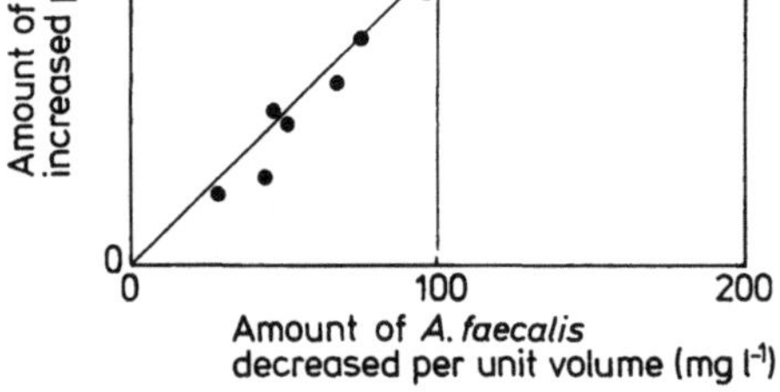

Fig. 7. Yield coefficient for conversion from *Alcaligenes faecalis* to *Vorticella microstoma* [30]

Table 7. Examples of maximum growth rate, saturation constant and yield coefficient for conversion from bacteria to Ciliata in batch culture

Predator	Prey	Temperature (C)	$\mu_{p,max}$ (d^{-1})	K_x ($mg\ l^{-1}$)	($N\ ml^{-1}$)	Yield	Ref.
Colpoda steinii	*Escherichia coli*	30	5.5	6.0	1.5×10^7	0.78	12)
Entodinium caudatum	*Escherichia coli*	—	—	—	—	0.50	13)
Tetrahymena pyriformis	*Klebsiella aerogenes*	25	5.3	11.6	—	0.50	14)
Tetrahymena pyriformis	*Aerobacter aerogenes*	25	2.4	6.1	—	0.73	15)
Vorticella microstoma	*Alcaligenes faecalis*	20	2.2	40	7.2×10^7	0.47	11)
Colpidium campylum	*Alcaligenes facalis*	20	2.9	11	2.0×10^7	0.53	3)
Opercularia sp.	*Alcaligenes faecalis*	20	2.3	36	6.5×10^7	—	3)

weight of protozoa is particularly susceptible to error, unless the cultivation proceeds axenically and is free from forming protozoa-bacteria aggregate.

An average dry weight of *Vorticella microstoma* and *Alcaligenes faecalis* obtained here was 3.85×10^{-6} mg per cell and 5.56×10^{-10} mg per cell, respectively [3]. The yield of *Vorticella microstoma* from *Alcaligenes faecalis* was 1.2×10^5 protozoa per mg bacteria (1.8×10^9 cells). It follows that the production of a single cell of *Vorticella microstoma* requires 1.5×10^4 cells of *Alcaligenes faecalis*. If the specific growth rate of *Vorticella microstoma* is 1.8 d^{-1} ($t_d = 9.1$ h), the protozoa would consume about 27 cells of *Alcaligenes faecalis* every minute during the interval between binary fissions.

The yield coefficient (ca. 0.5) for Ciliata production from bacteria indicates that the amount of sludge produced depends on whether protozoan growth occurs. According to observations and experiences of those who have been engaged in the operation of activated sludge processes of wastewater treatment, it has been widely recognized that the amount of excess sludge produced during summer when protozoa proliferate more actively is reduced in comparison with that of winter, during which the protozoan growth deteriorates [16]. Indeed, one of the authors (R.S.) has observed the tendency to produce less excess sludge in full-scale activated sludge plants when protozoa were prevalent, although this experience that is not necessarily new has yet to be confirmed in quantitative terms. To recapitulate, the protozoan role in biological wastewater treatment processes is considered to manifest itself both with respect to effluent quality and in decreases in excess sludge production when protozoa thrive in the plant.

4 Monoxenic and Continuous Culture

It has been claimed that biological treatment systems that discharge effluent of good quality usually contain large numbers of various protozoa. Protozoa play an important role in the improvement of effluent quality. Generally, the concentration of coliform bacteria in the effluent from an activated sludge process is low when Ciliata are the dominant protozoa in the sludge [11,17]. Further, the protozoa in the sludge tend to accelerate the flocculation of the activated sludge bacteria, thus facilitating gravitational separation of the sludge from the mixed liquor [18–20].

Such roles of the protozoa have only been demonstrated on a qualitative basis and the lack of quantitative information can be ascribed to both difficulties in the isolation of and in the monoxenic cultivation of protozoa. Hence, monoxenic and continuous cultivations of protozoa with sludge bacteria are required to establish a quantitative picture on the behavior of protozoa in the biotic community.

The relationship between predator and prey has been discussed from the ecological viewpoint by Gause [21] and Odum [22]. In connection with the biological treatment of wastewaters, several experimental works on monoxenic and continuous cultivation of protozoa have been published; e.g. predation of *Tetrahymena pyriformis* on *Escherichia coli* and *Azotobacter vinelandii* [23,24], *Dictyostelium discoideum* on *E. coli* [25], and *T. pyriformis* on *Aerobacter aerogenes* [15]. Initiated by Lotka and Volterra [26] as harbingers of the theoretical analysis of the predator-prey model, Bungay and Bungay [27], and more recently, Canale et al. [15,28,29] discussed theoretically the oscillation of predator-prey population densities for protozoa and bacteria in biological wastewater treatment processes. Although these works dealt with the ecological aspects of predator-prey relationship, assumption on the role of protozoa in the biological wastewater treatment systems were never justified by basic experiments.

In this section, monoxenic and continuous cultivation of the protozoon, *Colpidium campylum*, using the bacterium, *Alcaligenes faecalis*, and asparagine as the limiting substrate for the bacterium will be presented in order to augment our knowledge and to clarify the points that have been left unclear by the previous predator-prey experimental studies.

4.1 Batch Culture of *Colpidium campylum*

The value of the specific growth rate, μ for *C. campylum* cultivated in batch culture is plotted against the initial concentration of *Alcaligenes faecalis* as shown in Fig. 8 [3]. It can be seen that μ values can, for bacterial concentrations of less than 100 mg l^{-1}, be represented by the Monod equation (see a broken line). However, an increase of the bacterial concentration beyond a critical level (nearly 200–300 mg l^{-1} in Fig. 8) results in a decrease of the values of μ unlike the situation predicted by conventional Monod kinetics.

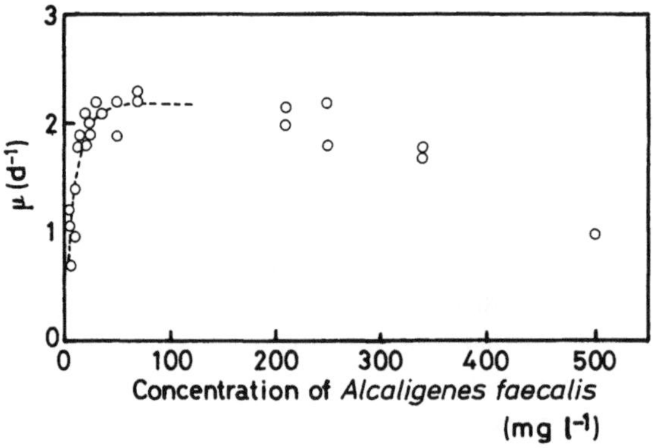

Fig. 8. Specific growth rate of *Colpidium campylum* (20 C) vs concentration of *Alcaligenes faecalis*

Although the values of μ decreased when the bacterial concentration exceeded a specific value, the number of bacterial cells that were entrained in a fluid stream and then, sucked into the cytostome of the protozoa tended to decrease when the bacterial population increased beyond the critical level [3]. The physical and micro-hydrodynamic behavior of the bacterial suspension around the protozoa might be responsible for the decrease in μ values in Fig. 8. Assuming the Monod-type saturation model,

$$\mu = \mu_{max} \frac{X}{K_x + X} \tag{1}$$

μ_{max} and K_x values could be estimated, respectively as follows:

$\mu_{max} = 0.121\ h^{-1}$
$K_x = 11\ mg\ l^{-1}$ $(2.0 \times 10^7$ cells per ml)

4.2 Chemostat Culture of *Alcaligenes feacalis*

Prior to the discussion of a mixed continuous culture of the protozoon and the bacterium, it is necessary to establish whether *Alcaligenes faecalis* can be grown in chemostat culture, and basic data on the maximum specific growth rate and the saturation constant of the bacterium must be secured. The results of the continuous cultivation of *A. faecalis* utilizing asparagine as carbon and nitrogen source are shown in Fig. 9 [30]. At each dilution rate, the cell population density (OD at 610 nm) and the residual asparagine concentration were measured intermittently. When each value was confirmed to remain almost unchanged after an initial working volume was replaced three to four times by the incoming fresh medium, the steady state was assumed.

The results in Fig. 9 clearly demonstrate that asparagine was the limiting substrate.

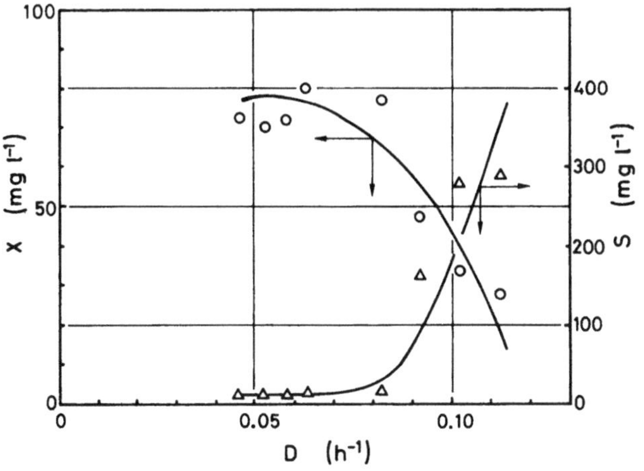

Fig. 9. Chemostat culture of *Alcaligenes faecalis* (limiting substrate: asparagine) [30]. Solid lines for X and S are drawn arbitrarily, not from calculation

A dilution rate, D, approaching $0.090 \, h^{-1}$ was most susceptible to variations. The dilution rate near $0.090 \, h^{-1}$ was far from "wash-out". Needless to say, a specific dilution rate, at which the wash-out occurs, was deemed more than $0.10 \, h^{-1}$ (Fig. 9). However, in the range of dilution rates from 0.046 to $0.064 \, h^{-1}$, a series of similar steady states were achieved. The value of yield factor, Y, for this range of dilution rates was from 0.148 to 0.164; an average was taken as $Y = 0.15$ for ease of the simulation that appears later on. The values of the maximum specific rate of growth μ_m, and saturation constant, K, for this bacterium were obtained from the Lineweaver-Burk plot. These are $\mu_m = 0.114 \, h^{-1}$ and $K = 10.7 \, mg \, l^{-1}$.

During the continuous culture experiments no bacterial flocs were observed, the bacterial cells being uniformly dispersed throughout the medium at all dilution rates in Fig. 9 (for the absence of protozoa). Indeed, the presence of protozoa can cause an appreciable level of bacterial flocculation as will be referred to next.

4.3 Mixed Continuous Culture

The results of a mixed continuous culture of *Colpidium campylum* and *Alcaligenes faecalis* at a dilution rate, D, of $0.065 \, h^{-1}$ are shown in Fig. 10. In this figure, the

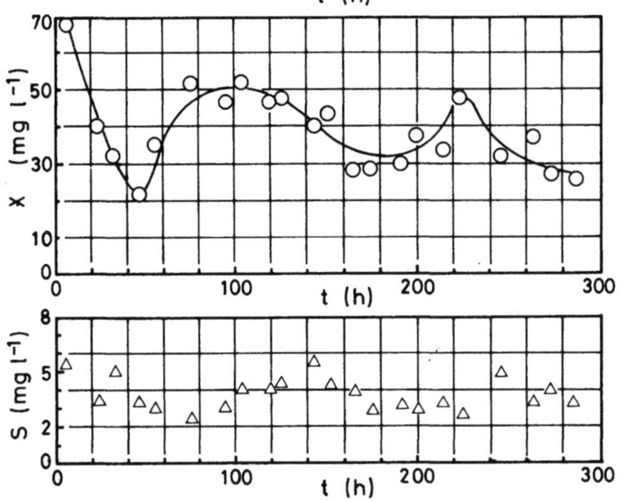

Fig. 10. Example of mixed continuous culture (*Colpidium campylum* and *Alcaligenes faecalis*), $D = 0.065 \, h^{-1}$ [30]

top, middle, and bottom diagrams deal with the population densities of *Colpidium campylum* (P), *Alcaligenes faecalis* (X), and the concentration (S) of residual asparagine in the effluent against time, t, respectively [30].

Fig. 10 shows that the population densities of the protozoon and bacterium oscillated, but the oscillatory behavior of the residual asparagine concentration was less evident in view of the initial concentration of asparagine, $S_0 = 500\ mg\ l^{-1}$. In addition, it was of interest to find that the maxima and minima of predator and prey densities were in the opposite sense. The periodicity was about 100 h. Similar oscillatory phenomena between the predator and prey were always observed in other experimental runs than Fig. 10. The predator and prey studies on activated sludge by Canale [28] come to similar results.

Intuitively, the bacteria population should have been exhausted with an increase in protozoan density. However, bacterial flocs began to appear when the protozoan density approached $4\ mg\ l^{-1}$ (2.5×10^3 per ml). When the protozoan population exceeded $10\ mg\ l^{-1}$ (6×10^3 per ml), swarms of considerably large flocs of bacteria (about 100 to 500 microns in diameter) were observed. It was most probable that the formation of these flocs prevented extermination of the bacterium and this situation gave rise to further proliferation of *Alcaligenes faecalis* in contrast to the protozoan decay which obviously originated from a lack of food. On the other hand, the bacterial flocs disintegrated with a decrease in protozoan density and the previous cycle between the predator and prey densities was repeated.

Curds and other workers also emphasized the protozoan role in the creation of bacterial flocs [18–20]. Their emphasis seems to have been supported by the observation (cf. Fig. 10) that initiation and/or termination of the floc formation depended on the protozoan population and its mucous material as a flocculating agent.

In contrast to the patterns for protozoa and bacteria in Fig. 10, the damped oscillatory behavior of asparagine might be ascribed to the possibility that *Colpidium campylum* metabolized some fraction of the dissolved nutrient, although at a much lower rate by *Alcaligenes faecalis*.

The possibility just mentioned could be envisaged from the fact that the residual concentration of asparagine, S, in the effluent from the chemostat culture of *Alcaligenes faecalis* was $S = 13.2\ mg\ l^{-1}$ at $D = 0.064\ h^{-1}$, far larger than that of $S(S = 2 \text{ to } 5\ mg\ l^{-1})$ for the mixed culture operating at nearly the same dilution rate (cf. Figs. 9 and 10).

Additional evidence of protozoan uptake of asparagine was derived under conditions where the bacteria were inactivated by adding penicillin (1,000 units per ml) and streptomycin ($100\ \mu g\ ml^{-1}$) to the mixed culture. Asparagine, $S_0 = 500\ mg\ l^{-1}$ in the fresh medium was consumed up to: $S_0 - S = 200$ to $50\ mg\ l^{-1}$ in 24 h. No protozoan growth (neither size nor number) was observed; its concentration was $(3 \text{ to } 7) \times 10^3$ cells per ml or 5 to $11\ mg\ l^{-1}$. If the consumption of asparagine by *Colpidium campylum* were assumed in the monoxenic culture with viable *Alcaligenes faecalis*, the residual concentration ($2 \sim 5\ mg\ l^{-1}$) of asparagine in Fig. 10 would be acceptable.

5 Mathematical Analysis of Mixed Culture

Mass balance equations for limiting substrate, S, bacterium, X and protozoa, P in a completely mixed continuous reactor are:

$$\frac{dS}{dt} = D(S_0 - S) - \frac{F(S, X)}{Y} \tag{2}$$

$$\frac{dX}{dt} = F(S, X) - DX - \frac{G(X, P)}{W} \tag{3}$$

$$\frac{dP}{dt} = G(X, P) - DP \tag{4}$$

Multiplying both sides of Eqs. (2) and (3) by YW and W, respectively and then, adding both sides of these equations and Eq. (4),

$$YW\frac{dS}{dt} + W\frac{dX}{dt} + \frac{dP}{dt} = DYWS_0 - D(YWS + WX + P) \tag{5}$$

Setting,

$$YWS + WX + P = \Phi$$

Eq. (5) can be rearranged, so:

$$\frac{d\Phi}{dt} + D\Phi = DYWS_0 \tag{6}$$

The linear differential equation can be solved to give:

$$\Phi = YWS_0 + Ce^{-Dt} \tag{7}$$

provided that C is an empirical constant.

If t becomes large, the second term on the right-hand side of Eq. (7) can be disregarded, and the following equation is obtained.

$$YWS + WX + P = YWS_0 \tag{8}$$

Since

$$Y \simeq 0.15\,,$$

$W \simeq 0.5$, [11,16]

$$S \lesssim 5\ \text{mg}\ l^{-1}\,,$$

$$X \simeq 10 \sim 50\ \text{mg}\ l^{-1}\,,$$

$$P \simeq 5 \sim 16\ \text{mg}\ l^{-1} \quad \text{and} \quad S_0 = 500\ \text{mg}\ l^{-1}\,,$$

(see above; cf. Fig. 10) the first term on the left-hand side of Eq. (8) can be neglected compared with the second and third terms without introducing any serious error. Accordingly,

$$WX + P = YWS_0 \tag{9}$$

Equation (9) represents a straight line on a P-X plane, because the term on the right-hand side of the equation is assumed constant (see Fig. 11a). Points A and B in Fig. 11a are singular points, since Point A corresponds to the maximum value of P (minimum value of X), whereas Point B refers to the contrary i.e., Points A and B satisfy simultaneously that

$$\frac{dP}{dt} = \frac{dX}{dt} = \frac{dS}{dt} = 0$$

in Eqs. (2) to (4). So far as the experimental data (Fig. 10 and other experimental data not shown here) are concerned, oscillations of P and X were observed along line AB in Fig. 11a, having two singular points for $P > 0$ and $X > 0$.

A classical model initially [26] presented by Lotka and Volterra and refined by Bungay and Bungay [27] is:

$$\frac{dX}{dt} = \mu X - DX - k_1 XP \tag{10}$$

$$\frac{dP}{dt} = k_2 XP - DP \tag{11}$$

The phase plane analysis shows that the populations of X and P oscillate around a vortex point as illustrated in Fig. 11b [28].

It is noted that the vortex point represented by $\tilde{P} = (\mu - D)/k_1$ and $\tilde{X} = D/k_2$ on the P-X plane is the one and only singular point for the first quadrant. As was referred to earlier, the experimental data suggest two singular points (cf. Fig. 11a). Consequently, the model shown by Eqs. (10) and (11) does not explain the experimental data.

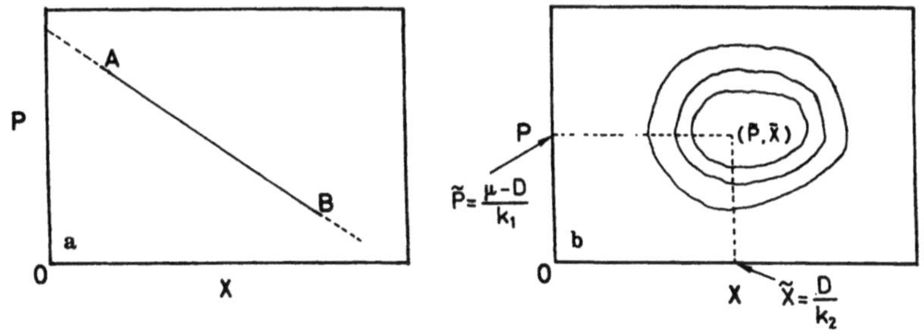

Fig. 11a. Relationship between P and X; **b** Solution of Lotka-Volterra model [28], parameter; initial condition

Recently, Canale [11] published another predator-prey model in which F(S, X) and G(X, P) appearing in Eqs. (2) to (4) were assumed as follows:

$$F(S, X) = \mu_m \frac{S}{K + S} X \tag{12}$$

$$G(X, P) = \mu_{p,\,max} \frac{X}{K_x + X} P \tag{13}$$

Although the emergence of a limit cycle on the phase plane is cited by Canale [29] as responsible for the oscillation, the experimental data on protozoan predation on bacteria by Canale et al. [15] cannot be attributed totally to the limit cycle. In fact, the model (Eqs. (2) to (4), (12), and (13)) could not simulate the data in Fig. 10 particularly when the smaller amplitude in the change of asparagine concentration with time was concerned (computation details are omitted).

These results, along with experiments that confirmed the emergence and/or disappearance of bacterial flocs depending on protozoan population, suggest that the bacterial population be divided into two categories; one, the bacterial flocs that are not eligible as food to protozoa, while the other category deals with dispersed bacteria which could become food for protozoa. Although the mechanism on bacterial flocculation and/or deflocculation incorporated into a new model [30] was rather intuitive and could not be confirmed by separate experiments, the model could simulate the non-uniform predator-prey population dynamics shown in Fig. 10.

It should be pointed out that characteristic constants (μ_m and K) of *A. faecalis* assessed separately (Fig. 9) could not be applied to the new model. Since observed values for S and X ranged approximately from 2.5 to 5.5 mg l^{-1}, and 20 to 60 mg l^{-1}, respectively in the experiment (Fig. 10), and also, since the values of dS/dt ($-0.2 \sim 0.2$ mg l^{-1} h^{-1}) could be estimated by smoothing the data points in Fig. 10, the values of F(S, X)/Y (Eq. (2) provided: $D = 0.064\ h^{-1}$ and $S_0 = 500$ mg l^{-1}) ranged as follows:

$$31 \leqq \frac{F(S, X)}{Y} \leqq 33 \tag{14}$$

The same term estimated from the separate experiment on *A. faecalis* (Fig. 8) using Eq. (12) and taking $\mu_m = 0.114\ h^{-1}$, $k = 10.7$ mg l^{-1} and $Y = 0.15$ is:

$$2.8 \leqq \frac{F(S, X)}{Y} = \frac{\mu_m}{Y} \frac{S}{K + S} X \leqq 16 \tag{15}$$

Eq. (15) indicates that the term F(S, X)/Y for *A faecalis* (pure culture) was far less than the value for the mixed culture. In fact, if the values of $\mu_m = 0.114\ h^{-1}$ and $K = 10.7$ mg l^{-1} in Fig. 8 were used in the model, the protozoan population became dominant, minimizing the oscillation in population densities; in other words, it was far from reality. The values actually used to simulate experimental results were $\mu_m = 0.3\ h^{-1}$ and $K = 5$ mg l^{-1} [30]. These values suggest the "activation" of the bacterial proliferation for which further investigations are required.

Arbitrary as this choice of values may sound, characteristic values of microbes, when measured individually, would not necessarily represent those when mixed. Experimental observations on ecological competition among different species of microbes point out this inconsistency between pure and mixed cultures [31]. Indeed, there remains ample room for further work on this aspect of mixed culture.

6 Conclusion

1) Bio-film processes can be subdivided into three categories: submerged bio-film, trickling filter and rotating discs according to differences in the oxidation. Of the three categories, the use of rotating discs in small-scale facilities to treat industrial wastewaters is now explanding. However, the submerged bio-film and especially, trickling filter, whose history of expansion dates back earlier than the activated sludge process, are still being used worldwide.
2) Microbial fauna in the activated sludge has been investigated. Ciliata might serve as an index for judging the quality of the effluent. The improvement of the effluent quality by certain Ciliata might be attributed to the following two factors: first, protozoan predation of bacteria and secondly, microbial flocs that would be liable to emerge in the presence of protozoa. Especially, microbial flocs facilitate the settling of the sludge, hence warranting an effluent of good quality from the sedimentation basin.
3) Some specific bacteria that would be eligible for food of the three species of *Vorticellidae* (*Vorticella microstoma. Vorticella convallaria* and *Carchesium polypinum*) were examined. Many genera and species of bacteria could support the proliferation of *V. microstoma.* This observation substantiates the frequency occuring dominance of *V. microstoma* in the activated sludge.
4) It is most probable that the *Vorticellidae* are holozonic, i.e. the protozoa metabolize only living bacteria, taking no soluble and/or insoluble organic substrates directly. The conversion efficiency of bacteria to protozoa was about 0.5 g protozoa per g bacteria. If the protozoa in the sludge were enhanced, the population enhancement would be accompanied by a considerable reduction of excess-sludge amount. The argument on excess-sludge reduction is justified, albeit not strictly, by a long experience in operating wastewater treatment.
5) Monoxenic and continuous culture of protozoa isolated from the activated sludge point out the ways and means how to keep a species of protozoa dominant in the sludge.
6) In mixed culture (monoxenic), the specific growth rate of protozoa as a hyperbolic function of the bacterial concentration has been confirmed. When the bacterial concentration is exceedingly high, the specific growth rate of protozoa deteriorates sharply. This phenomenon is explained by the protozoan micro-hydrodynamic characteristics.
7) Oscillatory phenomena among the population densities of protozoa, bacteria and soluble substrate have been confirmed experimentally, but the theoretical analysis needs to be improved, especially with kinetic constants involved.

7 Nomenclature

C	—	empirical constant
D	h^{-1} or d^{-1}	dilution rate
F	—	function of S and X
G	—	function of P and X
k_1, k_2	—	rate constants
K	$mg\,l^{-1}$	saturation constant for bacteria
K_x	$mg\,l^{-1}$	saturation constant for protozoa
P	$mg\,l^{-1}$	protozoan mass concentration
S	$mg\,l^{-1}$	substrate concentration
S_0	$mg\,l^{-1}$	substrate concentration in fresh medium
t	h	time
t_d	h	doubling time
W	—	yield factor of protozoan growth defined by $\Delta P/(-\Delta X)$
X	$mg\,l^{-1}$	bacterial cell mass concentration
Y	—	yield factor of bacterial growth defined by $\Delta X/(-\Delta S)$
μ	h^{-1}	specific growth rate
μ_m	h^{-1}	maximum value of μ
μ_p	h^{-1}	specific growth rate of protozoa
$\mu_{p,max}$	h^{-1}	maximum value of μ_p
Φ	—	$YWS + WX + P$
$\sim$	—	singular point (superscript)

8 References

1. Sudo, R., Ohgoshi, Y.: Seasonal changes of activated sludge microorganisms (in Japanese), Water Purification and Liquid Waste *5*, 43 (1964)
2. Pike, E. B., Curds, C. R.: in Soc. Appl. Bacteriol. Symp. Ser. 1. Microbial Aspects of Pollution. (Sykes, G., Skinner, F. A. eds.) p. 123, London Academic Press 1971
3. Sudo, R.: Studies on the role of protozoa in the activated sludge process. (in Japanese) Study on the Protozoan Role in Activated Sludge Processes, D. S. Thesis, Uni. of Tohoku, Japan 1974
4. Jones, G. L.: Nature *243*, 546 (1973)
5. Sudo, R., Aiba, S.: Jap. J. Ecol. *21*, 140 (1971)
6. Sudo, R., Aiba, S.: ibid. *21*, 70 (1971)
7. Sudo, R., Aiba, S.: in Proc. IV IFS: Ferment. Technol. Today, (Terui, G. ed.) p. 577, Soc. Ferm. Technol. Osaka 1972
8. Curds, C. R., Vandyke, J. M.: J. Appl. Ecol. *3*, 127 (1966)
9. Fineley, H. E., Mclaughlin, D., Farrison, D. M.: J. Protozool. *6*, 201 (1959)
10. Coler, R. A., Gunner, H. B.: Water Res. *3*, 149 (1969)
11. Sudo, R., Aiba, S.: ibid. *7*, 615 (1973)
12. Proper, G., Garver, J. C.: Biotech. Bioeng. *8*, 287 (1966)
13. Coleman, G. S.: J. Gen. Microbiol. *37*, 209 (1964)
14. Curds, C. R., Cockburn, A.: ibid. *54*, 343 (1968)
15. Canale, R. P., Lusting, T. D., Kehrberger, P. M., Salo, J. E.: Biotech. Bioeng. *15*, 707 (1973)
16. Sudo, R.: Biology for Wastewater Treatment (in Japanese), p. 281, Sangyo-Yosui Chosakai, Tokyo 1977
17. Curds, C. R., Fey, G. J.: Water Res. *3*, 853 (1969)
18. Curds, C. R.: J. Gen. Microbiol. *33*, 357 (1963)
19. Hardin, G.: Nature, Lond., *151*, 642 (1943)

20. Watson, J. M.: ibid. *155*, 271 (1945)
21. Gause, G. F.: The Struggle for Existence, Baltimore: Williams and Wilkins 1934
22. Odum, E. P.: Fundamentals of Ecology, Philadelphia: W. B. Saunders Co., 1971
23. Jost, J. L., Drake, J. F., Fredrickson, A. G., Tsuchiya, H. M.: J. Bacteriol. *113*, 834 (1973)
24. Bonomi, A., Fredrickson, A. G.: Biotechn. Bioeng. *18*, 239 (1976)
25. Tsuchiya, H. M., Drake, J. F., Jost, J. L., Fredrickson, A. G.: J. Bacteriol. *110*, 1147 (1972)
26. Lotka, A. J.: Elements of Physical Biology, Baltimore: Williams and Wilkins, 1925
27. Bungay, H. R., Bungay, M. L.: Advan. Appl. Microbiol. *10*, 269 (1968)
28. Canale, R. P.: Biotech. Bioeng. *12*, 353 (1970)
29. Canale, R. P.: ibid. *11*, 887 (1969)
30. Sudo, R., Kobayashi, K., Aiba, S.: ibid. *17*, 167 (1975)
31. Shindala, A., Bungay, H. R., Kreig, N. R., Culbert, K.: J. Bacteriol. *89*, 693 (1965)

Bei Fragen zur Produktsicherheit wenden Sie sich bitte an:
If you have any questions regarding product safety,
please contact:

Walter de Gruyter GmbH
Genthiner Straße 13
10785 Berlin
productsafety@degruyterbrill.com